# Integrated Mathematics

## Choices and Challenges

# Integrated Mathematics

## Choices and Challenges

Edited by Sue Ann McGraw
Lake Oswego High School (Retired)
Lake Oswego, Oregon

1906 Association Drive, Reston, VA 20191-1502
(703) 620-9840; (800) 235-7566; www.nctm.org


ISBN 0-87353-530-8



The publications of the National Council of Teachers of Mathematics present a variety of viewpoints. The views expressed or implied in this publication, unless otherwise noted, should not be interpreted as official positions of the Council.

Printed in the United States of America

# Contents

# Preface

THE FOCUS of *Integrated Mathematics: Choices and Challenges* is the teaching of meaningful, significant, and useful mathematics using an integrated approach. *Principles and Standards for School Mathematics*, published by the National Council of Teachers of Mathematics (NCTM) in 2000, strongly recommends integrating mathematical topics both in mathematics courses and in other subjects. This integrated approach, used internationally, enables students to view major mathematical ideas from more than one perspective as well as use a variety of tools and approaches in trying to understand new topics or solve problems.

In surveying its members in recent years, NCTM determined that the definition of integrated mathematics across grade levels lacked clarity and uniformity and that teachers had various concerns about whether their backgrounds and materials were adequate for teaching an integrated curriculum. In 1996, the NCTM president formed a Task Force on Integrated Mathematics to study and report on integrated mathematics in kindergarten through grade 12. The task force recommended to the NCTM Board of Directors that NCTM publish a book to assist educators and others in understanding the nature of teaching mathematics in an integrated approach and to guide the planning, implementing, and assessing of curricula for integrated mathematics. The Board accepted the recommendation, and hence this project was launched.

The resulting book, *Integrated Mathematics: Choices and Challenges*, is organized into four parts. Part 1, comprising chapters 1 through 6, addresses broad issues in, and perspectives on, integrated mathematics. Chapter 1 presents an overview of integrated mathematics as an ongoing issue, lists reasons to integrate programs, and addresses areas of concern. Chapter 2 offers a historical perspective on integrating the school mathematics curriculum in the United States, describes different kinds of integration within mathematics, and illustrates how differences in the size of the curriculum can affect the possibilities for integrating mathematics. Chapter 3 emphasizes both the importance and necessity of selecting significant mathematical concepts and ideas to teach. Chapter 4 presents an overview of general integration models, particularly those that integrate science and mathematics, and briefly describes the thirteen *Standards*-based mathematics curriculum programs that were developed with support from the National Science Foundation (NSF). Chapter 5 describes effective instruction that can help students understand how mathematical ideas interconnect and build on one another.

Chapter 6 illustrates how one Content Standard, data analysis, can be developed throughout the grade levels and how such integration can help students connect mathematical ideas as well as apply mathematics in other contexts.

Part 2, consisting of chapters 7 through 10, includes information on how to plan for, implement, and assess curricula for integrated mathematics. Chapter 7 outlines important conditions for any successful cross-disciplinary integration and describes how to develop worthwhile problems, adjust teaching strategies, and design appropriate assessment in the context of integrated mathematics and vocational-technical curricula. Chapter 8 presents a collaborative model for a college-level, integrated calculus and physics course and evaluates student outcomes. Chapter 9 presents theoretical models for integrating mathematics and science and examines how aspects of these models can be presented to preservice elementary school teachers in a college-level teacher preparation program. Chapter 10 details a senior-level college course on mathematical modeling that strives to prepare prospective secondary school teachers to use technology appropriately in teaching an integrated mathematics curriculum.

Part 3, which comprises chapters 11 through 16, covers six classroom examples of integrated mathematics. The classroom teachers who have written these chapters have (*a*) used national, state, or district standards to select grade-level-appropriate mathematics; (*b*) designed activities that are student friendly, promote understanding, and help students make connections; and (*c*) incorporated several forms of assessment to measure students' understanding. Chapters 11, 12, and 13 describe thematic units that integrate mathematics across the curriculum. Chapters 14 and 15 explore ways to use literature as a context for mathematics instruction and include advice on selecting appropriate books and using literature as an effective teaching tool. Chapter 16 presents two projects that integrate mathematics, science, and technology at the middle school level in Australia and analyzes the successful and not-so-successful student outcomes.

Part 4, composed of chapters 17 through 20, presents the challenges of, and implications for, teaching integrated mathematics and describes the respective roles and responsibilities of various stakeholders in successfully implementing an integrated curriculum. Chapter 17 describes how one state used various integrated curriculum materials to help teachers make better curriculum choices, prepare to teach a curriculum focused on problem solving, and adapt to newly redesigned state assessments. In chapter 18, a teacher reflects on the challenges and issues that she faced and continues to face in using an integrated approach. In chapter 19, a mathematics coordinator describes the experience of, and issues surrounding, districtwide implementation. Chapter 20 wraps up the book by addressing the overall challenges related to implementing integrated mathematics programs and examining the various roles and players involved.

The production of this book represents the efforts of many committed educators over several years. The Editorial Panel in particular developed the guidelines for this publication, reviewed all the submitted papers, selected the papers to be included, and guided the shape of the final publication. I wish to thank the panel members, listed below, for their support and efforts:

| | |
|---|---|
| Donna Berlin | Ohio State University, Columbus, Ohio |
| Anita Bowman | University of North Carolina at Greensboro |
| Jose Franco | |
| William Higginson | Queens University, Kingston, Ontario |
| Peggy House | Northern Michigan University, Marquette, Michigan |

The writers of this publication also deserve special thanks. In sharing their expertise and describing their experiences with integrated mathematics, the chapter authors offer readers information and insight about curriculum planning and effective mathematics integration.

It is the hope of the Panel that *Integrated Mathematics: Choices and Challenges* will help educators more accurately meet the Connections Standard put forth in *Principles and Standards for School Mathematics* (NCTM 2000, p. 64)

> **Connections Standard**
>
> Instructional programs from prekindergarten through grade 12 should enable all students to—
>
> - recognize and use connections among mathematical ideas;
> - understand how mathematical ideas interconnect and build on one another to produce a coherent whole;
> - recognize and apply mathematics in contexts outside of mathematics.

Furthermore, the Panel hopes that all those committed to improving mathematics education can use this book to understand and consider the educational possibilities of connecting mathematical ideas.

# Part 1

## Introduction to Integrated Mathematics Curricula

1

# Integrated Mathematics: An Introduction

Peggy A. House

A CENTURY has passed since E. H. Moore delivered his 1902 presidential address to the American Mathematical Society, in which he observed the following:

> Engineers tell us that in the schools algebra is taught in one water-tight compartment, geometry in another, and physics in another, and that the student learns to appreciate (if ever) only very late the absolutely close connection between these different subjects, and then, if he credits the fraternity of teachers with knowing the closeness of this relation, he blames them most heartily for their unaccountably stupid way of teaching him (Moore 1970, p. 248).

As Moore went on to advocate for reforms in school mathematics instruction that he believed would result in a more coherent organization of the subject, he cautioned that such changes would need to be accomplished through evolution, not revolution. One cannot help but wonder whether Moore ever imagined that a century later, educators would still be grappling with the question. This volume is evidence that they are.

Since the publication of *Curriculum and Evaluation Standards for School Mathematics* (NCTM 1989), with its identification of mathematical connections as one of four Process Standards spanning the curriculum, the topic of integrating mathematics curricula has received renewed attention, although it had long been a subject of interest to educators and the focus of numerous recommendations throughout the twentieth century and earlier. Today the quest to understand what is meant by an integrated curriculum continues as educators seek to make decisions about how and whether to integrate the mathematics curriculum.

## WHAT IS INTEGRATED MATHEMATICS?

MUCH HAS been written about mathematical connections, or integrated mathematics. Publications dealing with this topic include an NCTM Yearbook (House 1995), an issue of *Mathematics Education Dialogues* (NCTM 2001), and numerous articles in all the NCTM journals, not to mention publications from other organizations and individuals. Yet those who survey such publications will quickly conclude that the challenge to understand, much less implement, an integrated mathematics program begins with the most fundamental of all questions—what is integrated mathematics?

In 1996, the NCTM Board of Directors appointed a Task Force on Integrated Mathematics and charged its members to "think carefully about the situation of integrated mathematics K–12, provide the Board with some background information, [and] make a recommendation to the Board about the need for a publication" (Dickey et al. 1997, p. 2). This book is the outcome of the Task Force's recommendation.

As they undertook the challenge given to them, the Task Force quickly concluded, "Integrated mathematics seems to have many meanings and interpretations. [They] found discussions about integrated mathematics to be problematic because participants may each have in mind their own, sometimes different, definition or interpretation of what it means to 'integrate' mathematics. The lack of consensus concerning what integrated curriculum or mathematics means seems to be a long standing problem" (Dickey et al. 1997, pp. 3–4).

One observation that several of this book's chapter authors have made is that integration usually means different things to teachers at different grade levels. For example, elementary school teachers often approach integration through thematic units. In the middle grades, integration may involve teachers of various subjects collaborating to develop cross-disciplinary instruction around a central topic. In high school, integrated mathematics is likely to signify a curriculum in which topics from various branches of mathematics (algebra, geometry, etc.) are combined in a single course. One common strategy for understanding the different approaches to integration is to represent them along a continuum from no integration (completely isolated school subjects) to full integration (total blending of all curricula). (See, e.g., Sherry Meier's article "Curriculum Integration: Part One—What Are the Issues?" [2000].)

This book will not satisfy those who seek a strict definition of integrated mathematics, but it will confirm the diversity that continues to exist. In Chapter 2, author Zalman Usiskin discusses five different approaches to integration within mathematics and across the curriculum and illustrates some of the attempts to implement those approaches in the past. He also introduces the useful notion of "sizes of curriculum" and shows how integration takes on different characteristics when applied in individual problems, whole lessons, units, or courses, as well as within a single subject or across an entire school curriculum. Other chapter authors offer different perspectives on integration.

Despite such variation in interpreting the term *integrated mathematics*, one can make some generalizations about the characteristics of integrated programs. Lott and Reeves (1991) reported the findings from a national survey of state and district mathematics supervisors, college and university mathematics educators, and secondary school mathematics teachers and concluded the following:

> An integrated mathematics program is a holistic mathematics curriculum that—
>
> - consists of topics from a wide variety of mathematical fields and blends those topics to emphasize the connections and unity among those fields;
> - emphasizes the relationships among topics within mathematics as well as between mathematics and other disciplines;
> - each year, includes those topics at levels appropriate to students' abilities;
> - is problem centered and application based;
> - emphasizes problem solving and mathematical reasoning;
> - provides multiple contexts for students to learn mathematics concepts;
> - provides continual reinforcement of concepts through successively expanding treatment of those concepts;
> - makes appropriate use of technology.

This list of characteristics furnishes a context for considering large-scale integrated curricula and a framework within which smaller units of integrated mathematics can be placed.

# WHY INTEGRATE?

ALTHOUGH *Curriculum and Evaluation Standards* (NCTM 1989) may have reopened the debate over mathematical connections or integrated programs and although *Principles and Standards for School Mathematics* (NCTM 2000) has reaffirmed and continued it, Burkhardt (2001) points out that only in the United States does integrated mathematics appear to be an issue. According to Burkhardt, "Nowhere else in the world would people contemplate the idea of a year of algebra, a year of geometry, another year of algebra, and so on." The following advantages of integrated curricula are adapted from Burkhardt's discussion (2001):

- Integrated curricula build essential connections through active processing over an extended period that first consists of weeks as the curriculum points out fundamental links and then ultimately encompasses years as the concepts are used in solving problems across a variety of contexts.
- Integrated curricula help make mathematics more usable by making links with practical contexts that give students opportunities to use their mathematics successfully in increasingly challenging problems.
- Integrated curricula avoid long gaps in learning that result from "year-long chunks of one-flavored curriculum."
- Integrated curricula allow a balanced curriculum with the flexibility to include newer as well as traditional topics of mathematics and to foster problem solving that spans several aspects of mathematics.
- Integrated curricula support equity because different branches of mathematics, for example, algebra and geometry, favor different learning styles, so an entire school year of one branch puts some students at a greater disadvantage than does a more balanced curriculum that includes several areas of mathematics.

Believing that developing integrated programs was important for the United States, the National Science Foundation in the early 1990s selected and funded and directed thirteen curriculum projects—three elementary school, five middle school, and five high school projects—to develop integrated mathematics curricula that reflected the vision of the NCTM *Standards*. In chapter 4, Donna Berlin reports on those projects, all of which have produced curricular materials now commercially available to schools. At this time, however, their implementation in schools is still limited.

Various arguments have been put forth as reasons for not adopting integrated programs. Among them are concerns that (*a*) significant mathematics content may be omitted or trivialized, (*b*) basic skills may receive inadequate emphasis, (*c*) too few teachers are adequately prepared to teach integrated curricula, (*d*) students may not perform as well on standardized assessments, and (*e*) students will have difficulty transferring between schools with different curricula. No strong body of research exists either to support or refute such concerns, although anecdotal reports are available on both sides of every claim.

During their deliberations, the NCTM Task Force on Integrated Mathematics also uncovered a number of issues that relate to implementing integrated programs. They issued the following statement (Dickey et al.1997, p. 12) to summarize those concerns:

> Integrated mathematics has the potential for enhancing the scope and power of mathematics teaching and learning. It also, however, has the potential for undermining the coherence and thrust of the mathematics curriculum that addresses our goals for all students stated in the NCTM *Curriculum and Evaluation Standards [for School Mathematics]*. The use of manipulatives or technology, links to literature, writing assignments, cooperative learning activities, or real-world applications cannot be considered curriculum simply because they feature mathematics combined with some other subject, mathematical topic, or set of skills. Nor can such instructional activities be substituted for genuine mathematics. To qualify as worthwhile mathematical tasks, content, instruction, and assessment must all engage students in using important ideas in ways that promote progress toward gaining mathematical power. Consequently, programs that feature a great deal of integration of mathematics with other school subjects—even programs ostensibly built around mathematics as the core of the curriculum—do not necessarily create meaningful mathematics learning. Unless they are developed as plans for accomplishing major mathematics teaching and learning goals, such programs may focus on trivial or disconnected information.

## QUESTIONS FOR CONSIDERATION

WHEN CONSIDERING the implementation of integrated mathematics programs, three areas of concern identified by the NCTM Task Force require careful attention: curriculum, teacher professional development, and assessment.

According to the Task Force report, "Curriculum planning is a critical and often neglected part of mathematics integration. To insure . . . that important learning goals are addressed, to insure that skill development and practice are provided, those who integrate curriculum must carefully map the learning goals they plan to achieve. Effort must also be made to avoid trivializing the content, forcing it into places that it does not naturally belong, or contriving ways to blend things that are perhaps better not blended." The Task Force proposed the following list of questions for evaluating integrated curricula (adapted from Dickey et al. [1997, pp. 13–14]):

- Does important mathematics drive the integration?
- Has the curriculum been carefully planned using Standards-based guidelines that address significant and level-appropriate mathematics?
- Does the mathematics integration enhance the goals of the curriculum?
- Does the integrated curriculum or topic address the core mathematics concepts found in the NCTM Curriculum and Evaluation Standards?
- Does the teacher consider a broad curriculum plan when selecting a unit, lesson, or activity?
- Does the integration enhance students' understanding of important mathematical ideas and their value?
- Does the unit build on previous knowledge?
- Can the unit, lesson, or topic, or some aspect of each, be better approached without integration?

Integrated curricula must be taught by teachers with the necessary knowledge not just of mathematics but also of other disciplines in which mathematics is applied, and this requirement in turn has implications for teacher education and professional development. Questions developed by the NCTM Task Force focus on such issues as the following (Dickey et al. 1997. p. 14):

- Are teachers sufficiently knowledgeable about the interrelations of mathematics?
- Are teachers sufficiently knowledgeable about different disciplines and the significant mathematics that supports other content areas?
- Have prospective teachers experienced a curriculum that explores meaningful connections within mathematics and interrelationships among disciplines?

- Are prospective teachers exposed to appropriate methods of multidisciplinary models of teaching?
- Do teachers have the freedom to make decisions about the overall curriculum?
- Does the teacher's workday allow for the extra planning time that integration demands?
- Does departmentalization in middle and high schools work against team planning and collaboration?
- Do developers of integrated curriculum offer assistance in curriculum mapping, staff development, and appropriate assessment strategies?

The third area of concern identified by the Task Force addresses assessment. Because most widely used assessment instruments are aligned with traditional subject-area divisions, some educators fear that students who study in integrated programs may be disadvantaged by nonintegrated tests. The Task Force generated the following questions with regard to assessment concerns (adapted from Dickey et al. [1997, p. 15]):

- Does the assessment of learning through an integrated approach require different assessment strategies or tests?
- What types of tests are sensitive to integrated approaches?
- Does testing integrated learning require a more subjective approach, and would this approach be accepted as a part of high-stakes testing?
- Are teachers who use an integrated approach penalized because tests are not sensitive to this approach?
- Does an integrated approach serve students whose achievement will be measured by discipline-specific tests?

## CONTINUING THE QUEST

CLEARLY, THE process of developing and implementing integrated mathematics programs is complex and challenging. The writers of the chapters in this book offer insights into some aspects of that complex challenge on the basis of their experiences with integration in varied contexts and at different levels of schooling. Their perspectives enable educators to further consider integrated mathematics as they strive to weigh instructional options and offer students the best possible learning opportunities.

The Curriculum Principle presented in *Principles and Standards for School Mathematics* calls for a mathematics curriculum that is coherent, focused on important mathematics, and well articulated across the grades. Such a curriculum will of necessity be "connected" and "integrated." And as we continue to search for deeper understanding and develop effective pedagogical alternatives, we can be both motivated and guided by the following vision of mathematics instruction from the introduction of *Principles and Standards* (NCTM 2000, p. 3):

> Imagine a classroom, a school, or a school district where all students have access to high-quality, engaging mathematics instruction. There are ambitious expectations for all, with accommodation for those who need it. Knowledgeable teachers have adequate resources to support their work and are continually growing as professionals. The curriculum is mathematically rich, offering students opportunities to learn important mathematical concepts and procedures with understanding. Technology is an essential component of the environment. Students confidently engage in complex mathematical tasks chosen carefully by teachers. They draw on knowledge from a wide variety of mathematical topics, sometimes approaching the same problem from different mathematical perspectives or representing the mathematics in different ways until they find methods that enable them to make progress. Teachers help students make, refine, and explore conjectures on the basis of evidence and use a variety of reasoning and proof techniques to confirm or disprove those conjectures. Students are flexible and resourceful problem solvers. Alone or in groups and with access to technology, they work productively and reflectively, with the skilled guidance of their teachers. Orally and in writing, students communicate their ideas and results effectively. They value mathematics and engage actively in learning it.

Imagine, indeed.

## REFERENCES

Burkhardt, Hugh. "The Emperor's Old Clothes, or How the World Sees It ..." *Mathematics Education Dialogues,* January 2001. www.nctm.org/dialogues/2001-01/20010102.htm (14 July 2002)

Dickey, Ed, Chris Brueningsen, Susan Butsch, Cindy Chapman, Miriam Leiva, and Harry Tunis. "Task Force on Integrated Mathematics Report to the

Board of Directors, National Council of Teachers of Mathematics." National Council of Teachers of Mathematics, Reston, Va., 1997. Photocopy.

Lott, Johnny W., and Charles A. Reeves. "The Integrated Mathematics Project." *Mathematics Teacher* 84 (April 1991):

Meier, Sherry L. "Curriculum Integration: Part One—What Are the Issues?" *NCSM Journal of Mathematics Education Leadership* 4 (Spring 2000): 2–8.

Moore, Eliakim Hastings. "On the Foundations of Mathematics." In *Readings in the History of Mathematics Education,* edited by James K. Bidwell and Robert G. Clason, pp. 246–55. Washington, D.C.: National Council of Teachers of Mathematics, 1970. Originally given as a presidential address before the American Mathematical Society (29 December 1902).

National Council of Teachers of Mathematics (NCTM). *Connecting Mathematics across the Curriculum.* 1995 Yearbook of the National Council of Teachers of Mathematics, edited by Peggy A. House. Reston, Va.: NCTM, 1995.

———. *Curriculum and Evaluation Standards for School Mathematics.* Reston, Va.: NCTM, 1989.

———. *Mathematics Education Dialogues* (entire issue). January 2001. www.nctm.org/dialogues/2001-01/20010105.htm (14 July 2002).

———. *Principles and Standards for School Mathematics.* Reston, Va.: NCTM, 2000.

2

# The Integration of the School Mathematics Curriculum in the United States: History and Meaning

Zalman Usiskin

IN EDUCATION, *integration* means the simultaneous consideration of different aspects of knowledge. In the United States, the question of the integration of the mathematics curriculum is often treated as many other issues in education are—with a simplistic view of the idea itself, discussed as if there is little or no historical context, and argued with varying combinations of myth and reality concerning its track record and promise. Despite the recommendations for integrated curricula found in some recent documents, very little analysis of the integration issue has occurred within the mathematics education community in the United States.

The situation in Canada is more complex because provinces are independent with respect to curriculum, and integration has been in place for some time in many areas. Yet, clarification of the issues can be useful even where an integrated curriculum is in place. As long as thirty years ago, Jim Hrabi, the provincial director of mathematics in Alberta, remarked, "The integration of the various fields of mathematics is a problem, and at the high school level the programme 'more closely resembles a mixture than it does a compound.'" (Hrabi 1967, as cited in Crawford 1970, p. 433).

This chapter attempts to provide a view of the complexity of the issue of the integration of the school mathematics curriculum in the United States. It does not discuss the situation in Canada. The chapter begins by placing integrated curricula in the context of the history of integration in mathematics. This overview leads to a description of different types of curricula that have been called integrated curricula. To clarify that discussion, the concept of size of curriculum is introduced. Finally, with these clarifying factors in place, the chapter considers what is and what is not an integrated curriculum and the accuracy of some of the explicit and implicit claims that have been made for integrated curricula. Throughout, the chapter presents numerous examples of and from integrated curricula.

## THE ORIGIN OF INTEGRATION

RENÉ DESCARTES began his essay *Géométrie* ([1637], 1954) by noting that lines and circles can be described algebraically and showing how their intersections lead to equations whose roots can be obtained by solving equations. He writes (p. 17), "These same roots can be found by many other methods, I have given these very simple ones to show that it is possible to construct all the problems of ordinary geometry by doing no more than the little covered in the four figures that I have explained." The major reason that Descartes could have such optimism was the new use of algebraic notation, less than forty years before, which had led to the development of analytic geometry by Fermat and Descartes himself. Since Descartes had already noted the long-established fact that the operations of arithmetic could be represented geometrically, the quote above carries with it his belief that every mathematical problem was solvable.

In Europe around Descartes's time, geometry was the paramount subject in the education of those few who became mathematicians, and it referred to a subject broader than what we think of as geometry today. Education in geometry came from Euclid's *Elements* (Heath 1956). Its thirteen books (which today we would call long chapters) contain far more than the geometry students typically encounter today. In addition to virtually every theorem in today's geometry textbooks and many more theorems not usually mentioned, the *Elements* contain a great deal of solid geometry no longer in the curriculum, as well as geometric equivalents of other results, including the law of cosines and the quadratic formula. The books also contain quite a bit of number theory, including the proof, done geometrically, that there are infinitely many primes. Although each book of the *Elements* is devoted to a particular mathematical area, the theorems in one book are used in other books as needed and the mathematical system is common throughout. The *Elements* is a text that integrated virtually all the mathematics known at the time of its writing.

In the seventeenth century, algebra had not yet developed into a separate discipline and analysis was treated geometrically. Algebra was merely a tool for Descartes and for other mathematicians. Even when Newton developed calculus in the 1680s, about fifty years after Descartes's work, he used geometric arguments rather than algebraic ones. These mathematicians gave no thought to whether mathematics was integrated, because there was only one mathematics, and it all developed out of geometry.

In the eighteenth century, during the Age of Enlightenment, the power of algebra and calculus together to solve large classes of problems gave some validity to Descartes's optimistic view of the power of mathematics. The importance of algebra and calculus began to overtake that of geometry. The major codification of algebra into a subject worthy of study on its own was by Euler. His *Elements of Algebra* (1770) was so influential that we owe much of today's algebraic notation, including the symbols $\pi$, *sin*, and $f(x)$, to their appearance in this text.

At first, the discovery of non-Euclidean geometries by Lobachevsky and Bolyai in the first half of the nineteenth century served to lessen the faith that mathematicians had in Euclidean geometry. Because for 2000 years no mathematician had realized that other geometries are possible, much of the mathematical community turned away from relying on geometry as the paragon of logical virtue and began turning attention and tracing arguments back to the foundations of arithmetic, analysis, and logic.

But three series of events in the nineteenth and early twentieth centuries served to raise algebra to at least a coequal place with geometry and ultimately led to a greater belief that mathematics is a unified whole. The first set of events began in the first third of the nineteenth century with the proof that solutions to polynomial equations of degree 5 and higher could not be expressed in a finite number of additions, multiplications, or powers and roots. The theory, due to Abel and Galois, did not employ geometry but used the algebraic structures of groups and fields. Then in 1872, Felix Klein used groups to classify all the geometries known until that time, including the non-Euclidean geometries. Thus, an idea undoubtedly algebraic in origin was employed to clarify the study of geometry.

The second series of events was the resolutions of the three famous unsolved construction problems of the ancient Greek mathematicians: (1) the construction of an angle with a measure one-third that of a given angle (trisecting an angle); (2) the construction of a square with area equal to that of a given circle (squaring the circle); and (3) the construction of a cube with volume twice that of a given cube (duplicating the cube). Working from algebraic theory rather than geometric relationships, each of these straightedge-and-compass constructions was proved impossible. These resolutions forcefully demonstrated that sometimes a significant theorem in one branch of mathematics can be proved or a significant problem can be solved only by using ideas, techniques, and methods from another branch.

The third set of events occurred at the beginning of the twentieth century. In 1899, the German mathematician David Hilbert showed that geometry was logically consistent if we assumed that arithmetic was logically consistent (Hilbert 1902), and within ten years Bertrand Russell and Alfred North Whitehead supplied details showing that arithmetic, algebra, geometry, and analysis could be viewed as emanating deductively from a common origin in logic.

## DIFFERENT KINDS OF INTEGRATION

THE TERM *integrated mathematics* has been applied to a number of different ways in which different areas of knowledge can be brought together. Five of these ways are described below: using unifying concepts, merging different areas of mathematics into broader areas, removing distinctions entirely between areas of mathematics, teaching different strands of mathematics each year, and interdisciplinary integration of mathematics with other subjects.

### Integration through Unifying Concepts

The many mathematical results that employ an idea in one branch of mathematics to understand or gain information about another branch give rise to the idea of a *unifying concept*—that is, a concept that cuts across most, if not all, branches of mathematics. Deduction is the oldest example of a unifying concept. From Euclid through Russell and Whitehead to the present day, mathematicians have utilized established principles of deduction. Teaching deduction and other logical principles has been considered a practice applicable not only in all of mathematics but also in life more generally.

The work of logicians in the early twentieth century established that set theory formed a basis for logic itself. A logical (pun intended) consequence was that when mathematicians looked to reconstruct the school curriculum in the "new math" era of the late 1950s and 1960s, *set theory* emerged as a unifying thread at virtually every level of mathematics instruction. The largest of the new math projects, the School Mathematics Study Group (SMSG), as well as other new math curricula, used such unifying concepts as sets, mathematical systems, and logic throughout their materials (NCTM 1961). The teaching of sets became so widespread in new math materials that, more than any other single topic, sets were viewed as a signal that a particular curriculum was developed in the spirit of the new math movement.

At the same time, in some European curricula designed for all students, *algebraic structures* gained prominence as a unifying concept because such properties as associativity, commutativity, and identity and inverse elements are exhibited in operations on numbers, matrices, transformations, vectors, and functions. France and French-speaking Belgium were highly influenced by a group of French mathematicians who, under the pseudonym Bourbaki, wrote for professional mathematicians a series of seminal books emphasizing general properties of sets, functions, and algebraic structures (Choquet 1964; Dienes and Golding 1967; Papy and Papy 1968). Even in schoolbooks, the mathematics was developed from these properties. The European ideas in turn influenced U.S. curricula designed for better students developed by the Secondary School Mathematics Curriculum Improvement Study (SSMCIS) (Fehr, Fey, and Hill 1972) and by the Comprehensive School Mathematics Project (CSMP) (*Elements of Mathematics* 1975; Krist 1985). In addition to showing off the structural unity of mathematics, teaching through these structures was viewed as an efficient and effective way to learn a good deal of mathematics.

More recently, we have seen the resurgence of a unifying concept that had prominence as a unifier earlier in this century—the concept of *functions*. Greater attention to functions in all parts of the curriculum was recommended in the influential 1923 report of a committee of mathematicians and teachers under the auspices of the Mathematical Association of America (National Committee on Mathematical Requirements 1923). A report written in the late 1930s considered functions to be a fundamental concept in problem solving, parallel to such processes as approximation, formulation and solution, and proof (Committee on the Function of Mathematics in General Education 1940). In the new math curricula of the 1960s, functions were used as a concept to unify the material within algebra and trigonometry. The recent interest in functions has taken advantage of the availability of automatic function graphers for older students and the interest in having younger students develop and describe relationships between and among variables.

This is not an exhaustive list of concepts that have played roles as unifying themes. In its *Agenda for Action*, NCTM (1980) recommended that the curriculum be organized around *problem solving*. *Algorithms* have been touted as a theme that transcends content and grade level (NCTM 1998). Often, to promote a unifying theme, a mathematical area is extended to become a way of thinking; for example, *algorithmic thinking* generalizes the ideas of algorithms; algebra may not be seen as a unifying concept but *algebraic thinking* is; functions might not be viewed as accessible to all students but *functional thinking* is recommended; and so on.

## INTEGRATION BY MERGING AREAS OF MATHEMATICS

IN A curriculum integrated through unifying concepts, traditional areas of school mathematics—such as arithmetic, algebra, geometry, analysis, and probability and statistics—may remain apart. A different type of integration, merging areas of mathematics, more closely follows the idea that integration is the bringing together of things that have been segregated or separated.

Before the 1960s, a student did not learn mathematics in grades K–6; rather, the student learned arithmetic—with textbook series having such titles as *Seeing Through Arithmetic* (Hartung et al. 1955). With new math came the push to teach some geometry, algebra, probability, set theory, and logic in these grades, and textbook titles changed to include the word *mathematics.* Although the "taught curriculum" of the 1960s was for many teachers still almost exclusively arithmetic, the "textbook curriculum" changed dramatically. Critics said that elementary school teachers were not ready for the new math, but without these developments, the more-recent recommendations for K–6 teachers to teach from all the branches of mathematics (NCSM 1977; NCTM 1980, 2000) would have little chance at implementation.

Before the 1960s, the U.S. high school curriculum for college-aspiring students was similarly devoted to individual branches. Algebra courses contained little or no geometry, not even as a context for solving equations. Geometry textbooks contained no number lines and no coordinate geometry. Except perhaps for the introduction of the right triangle ratios, trigonometry was taught in its own separate course. Solid geometry was taught in a course separate from plane geometry. In the spirit of Euclid, the course included work from what had been taught in plane geometry but contained no algebraic theory and no coordinates. Analytic geometry was a college course, and few of its ideas were seen earlier.

SMSG not only used unifying concepts but also broke down the barriers between courses. Solid geometry was merged with plane geometry into a single course that used number lines and distance functions and included coordinate geometry. Trigonometry was merged into the second-year algebra course (the name of which was changed by SMSG to *Intermediate Mathematics*) and into the study of functions, and the approach became more algebraic than geometric. Analytic geometry almost disappeared as a separate course, its topics being strewn among earlier courses (e.g., slope in first-year algebra, parabolas in *Intermediate*

*Mathematics*) or deleted altogether (e.g., the study of the cardioid curve, rose curves, and other special curves) (NCTM 1961, pp. 65–67). Almost all these major changes have become part of the standard secondary school mathematics curriculum and constitute major counterexamples to the thinking that the new math movement had no lasting influence. As a result of this influence, mathematics as currently taught tends not to be as compartmentalized as in the past.

The first of the current wave of reform projects, the University of Chicago School Mathematics Project, has developed a curriculum in which some courses maintain traditional titles and in which clear mathematical themes appear in each course—but the mathematics inside each course is unified through connections with other mathematics and with applications. The first course, *Transition Mathematics*, weaves arithmetic, algebra, and geometry throughout with no barriers at all. A significant amount of geometry and probability and statistics is in the *Algebra* and *Advanced Algebra* courses. Through uses of coordinates and transformations as well as of equation-solving and formulas, algebra and functions are themes throughout the *Geometry* course. The diverse mathematical areas identified in the titles of the last two courses in this curriculum, *Functions, Statistics, and Trigonometry* and *Precalculus and Discrete Mathematics*, are found not only in neighboring lessons and chapters but at times as applications of the same ideas (Usiskin 1993; Hirschhorn et al. 1995).

## Integration by Removing Distinctions between Areas of Mathematics

Merging areas of mathematics reaches its pinnacle when all mathematics is merged. The idea of breaking down all notions of teaching a particular area of mathematics has had proponents throughout the twentieth century. In 1901, E. H. Moore, the mathematics department chair at the University of Chicago, recommended that the traditional subject areas of school mathematics—algebra, geometry, and trigonometry—be unified. In the first decades of this century, a unified curriculum was developed first by George W. Myers and later by Ernest Breslich at the Laboratory Schools at the University of Chicago (Myers 1911; Breslich 1915, 1916, 1917; Breslich et al. 1916; Senk 1981).

Since the 1960s, the School Mathematics Project (SMP) and other mathematics projects in England have developed curricula in which topics from many different branches of mathematics—probability and statistics, geometry, algebra, and functions—are found in almost every year. Unifying ideas, such as set and logic and transformation, are found throughout, as are real-world applications.

Beginning in the early 1990s, with support from the National Science Foundation, a number of projects have developed materials for grades 9–12 in which the mathematics is developed around broad problem situations. The ARISE curriculum (Consortium on Mathematics and Its Applications [COMAP] 1998) is organized around units or modules based on such themes as fairness and apportionment, Landsat, and animation, with each module designed to provide about a month's worth of work in a school year. The Interactive Mathematics Program (IMP) curriculum (Alper et al. 1997) is built around problem units of five to eight weeks in length whose titles almost completely disguise the mathematics in them. For instance, of the five units in the first course, "Patterns," "The Game of Pig," "Overland Trail," "The Pit and the Pendulum," and "Shadows," only the first and last hint at the mathematics involved. The curriculum of the Systematic Initiative for Montana Mathematics and Science (SIMMS 1996) is similar, except that its units are shorter, being about two to two-and-one-half weeks in duration.

These recent curricula exhibit unifying concepts from applied mathematics, including, most important, mathematical modeling, and also simulation, estimation and approximation. The idea that applied mathematics might supply unifying concepts for school mathematics curricula is relatively recent, but the unity between mathematics and its applications has long been recognized. The greatest mathematicians of all time, from Archimedes through Newton and Euler to Gauss, worked in both pure and applied mathematics and back and forth from one to the other.

## Integration by Strands: Together but Separate

The most common form of integrated curriculum worldwide is teaching many areas of mathematics in the same school year while keeping their separate identities and without much concern for relating them.

For decades in the United States, textbooks from first through eighth grade or the year before algebra have contained chapters on geometry and measurement and on probability and statistics. So, if a curriculum consisting of strands or topics recurring year after year is considered to be an integrated curriculum, then most elementary school curricula throughout the world, including the traditional curricula most widely available in the United States, are integrated.

But if we consider a curriculum with strands to be integrated, then we must recognize that a common form of integrated curriculum in other countries of the world is to teach only two or three strands of mathematics each year and not relate these areas in anything more than superficial ways. In the mathematics curricula of Russia and the People's Republic of China, algebra and geometry are taught separately but simultaneously. In seventh and eighth grade, algebra is taught on certain days of the week and geometry on the other days, and each course uses a different book. Each course is self-contained with its own logical system, and few connections are made. This division continues through the early senior high school years. In Japan, the middle school textbooks in use through the 1980s and early 1990s included both algebra and geometry, but the subjects were quite separated. For instance, in one eighth-grade textbook (Kodaira 1984), the first four chapters are devoted to algebra, the next three are devoted to geometry, and the last is on statistics. In the current seventh-grade textbook used in Singapore, chapters 1–6 and 11–12 are arithmetic; chapters 7–8 are algebra, and the other chapters are geometry. The current Singapore eighth-grade textbook has more strands: chapters 1–2 are arithmetic, chapters 3–8 are algebra, chapters 9–11 and 13 are geometry, chapter 12 is statistics, and chapter 14 is trigonometry (Lee, Keng, and Chin 1997, 1998).

Starting in the late 1970s and continuing for about twenty years, the New York State Regents promoted a curriculum variously described as sequential, unified, or topical, based somewhat on the SSMCIS curriculum. Each year included some algebra, geometry, probability or statistics, and logic (Paul and Richbart 1985; Bumby and Klutch 1978–80; Rising et al., *Unified Mathematics*, Book 1, 1981; Rising et al., *Unified Mathematics*, Book 2, 1981; Rising et al., *Unified Mathematics*, Book 3, 1981). We might call this kind of integration "together but separate" because the different areas are together on paper but separated in actual teaching. In the words of Hrabi, quoted on the first page of this chapter, most curricula with strands are mixtures rather than compounds.

However, some curricula are compounds because the strands are connected in many ways. One such curriculum has been developed by the Middle Grades Mathematics Project for students in grades 6–8 (Lappan et al. 1998). The project identifies units on number, geometry, measurement, algebra, statistics, and probability, but many connections are made between strands, between mathematics and other disciplines, and between mathematical ideas and the real world. At the high school level, a similarly multiply connected curriculum is that of the Core-Plus Mathematics Project (Coxford et al. 1997). According to Core-Plus (1999. p. 4), "Each year the curriculum features strands of algebra and functions, statistics and probability, geometry and trigonometry, and discrete mathematics connected by fundamental themes, by common topics, and by habits of mind or ways of thinking."

## Interdisciplinary Integration

From the standpoint of integrating curricula, we note with interest that the oldest journal in the United States dealing with science education is *School Science and Mathematics*, which began in 1901 with the title *School Science*. In 1903, George W. Myers, a professor at the University of Chicago (mentioned above as the developer of a unified mathematics curriculum in the University High School) was asked to be the departmental editor for Mathematics and Astronomy for the Mathematical Supplement to *School Science*. Ultimately, this supplement became integrated into the journal and the title was changed to what it is today. Over the years, the journal became better known than the organization that sponsored it, the Central Association of Science and Mathematics Teachers. So in the 1970s, the organization changed its name to the School Science and Mathematics Association (SSMA).

The SSMA represents a view among science and mathematics educators that science and mathematics need at least to be considered, if not taught, together. The integration of science and mathematics was the subject of a 1991 national conference (Berlin and White 1992). The January 1994 issue of *School Science and Mathematics* was devoted to a discussion of the issues of integration of science and mathematics. Many of the suggestions of the American Association for the Advancement of Science Project 2061 involve teaching mathematics and science together, particularly at the elementary school level (Rutherford and Ahlgren 1990).

A few years ago, I presented some views on the integration of mathematics with science (Usiskin 1997). I argued that connections between mathematics and science are important, but so too are connections between mathematics and social studies. Also, in the elementary school, mathematics should be linked closely with reading. But I believe that mathematics should not be integrated with those other areas. It is too difficult to give enough attention to the different concepts in different subjects simultaneously, yet demonstrate the differences in their importance, and also link them in some coherent way.

Some educators have suggested merging more than two subjects. This movement has been strong particularly among educators at the middle school level but also has proponents at higher levels, such as the National Center for Cross-Disciplinary Teaching and Learning of the College Board (see also Fiscella and Kimmel 1999). The arguments given for joining all subjects or treating many subjects together parallel those given for integration within mathematics—

- knowledge is not inherently divided into categories, and our separation of learning by disciplines is artificial;

- the problems of today, such as pollution, population growth, and utilization of resources, are multidisciplinary; and
- problems are often handled in government, business, and industry by bringing together teams of people in different disciplines.

## THE SIZES OF CURRICULUM

TO FURTHER analyze the different kinds of integration, it is useful to introduce the idea of the sizes of curriculum.

The smallest size of curriculum is (1) the individual question, problem, or episode in teaching. This bit of curriculum may take from a few minutes to perhaps an hour of student time. Next in size is (2) the lesson, or problem set. The lesson is a set of questions or episodes and thus is an order of magnitude larger than the individual question or episode. A lesson typically takes a day or two of class time. A set of lessons or problem sets constitutes (3) a unit or, in typical books, a chapter. A unit typically includes six to fifteen lessons and may last from two to six weeks. A set of six to fifteen units constitutes (4) a course, typically a semester or year in length. Unlike smaller-sized curricula, courses yield grades that in later years of schooling are recorded and appear on transcripts. The set of mathematics courses taken or offered over a student's years of schooling constitutes (5) the school mathematics curriculum. We may think of this curriculum as being twelve to thirteen years long and split it into parts, each of which is a level of schooling. The set of curricula in the various areas of schooling constitutes (6) the school curriculum.

Each of these six sizes of curriculum is an order of magnitude larger than the previous. This fundamental difference causes significant differences in many other respects. For instance, a teacher typically has control over the materials used for sizes of curriculum smaller than the course, but a school or school district controls the materials for a course and for the curricula in all subjects. The evaluation of an individual question or episode is typically informal and done by the teacher as a class is being conducted. The evaluation of a lesson is slightly more formal, often by a short quiz that may be unannounced. These evaluations contrast with the evaluation of a unit by a longer teacher-made test for which preparation is expected, or with the evaluation of a course by a special school-constructed final examination for which more preparation is expected, and with the evaluation of the full curriculum by a standardized test constructed outside the school district and for which preparation is believed to have little effect.

## WHEN IS A CURRICULUM INTEGRATED?

THE DIFFERENCES in size of curriculum are reflected in the different possibilities for integrating mathematics. At the various sizes of the curriculum, integration means different things. A curriculum may be integrated when analyzed at one size and not at all integrated when analyzed at another.

Integration in individual problems means—

- to offer a wide variety of problems at all times, not sticking to problems of one type only;
- to allow the student all available means of solution on any individual problem (except when a particular strategy is being requested for purposes of analysis or practice of that strategy); and
- to apply the concept at hand to an idea from elsewhere in mathematics or to another discipline.

The student who is not allowed to use calculators is denied a chance at integrating technology and mathematics. The student who is not allowed to use arithmetic to solve an algebra word problem has lost an opportunity to integrate arithmetic with algebra. The application of an area formula to obtain the cost of carpeting a room has integrated mathematics with real life. The use of coordinates to prove that the length of a median of a trapezoid is the average of the length of the bases of the trapezoid is an integration of geometry with algebra.

An integrated lesson means that—

- one episode or problem flows to the next, and a mathematical content or process theme is dealt with in more than one way;
- a multitask activity connects mathematics with another subject;
- an idea is explained from a variety of mathematical viewpoints; or
- a wide variety of problems are used that involve the same concept but in different contexts.

Notice that an integrated lesson may have individual problems that are themselves not integrated, and also that one might have integrated problems that are not part of an integrated lesson.

Lessons that involve more than one subject area are far easier to plan than units involving more than one subject, so integrated lessons occur more often than integrated units, but they still are not common. When integrated units occur, it is probably more common for mathematics teachers to reach into other areas to bring people into their classes than the other way around. A writing assignment in mathematics may be examined by the English or language arts teacher. A science teacher may be invited into the mathematics classroom to perform an experiment. The music teacher may present sounds and their frequencies. And so on.

In an integrated unit, a unifying theme flows throughout the unit, somewhat like in the integrated lesson. Ideas introduced early in the unit are used later in the unit. Chapters in most textbooks, whether traditional or innovative, are integrated in this sense. When all subjects are involved in an integrated unit, the unit tends to be organized around a problem theme in contexts, such as the environment, the community, or a school event or set of holidays.

In an integrated course, connections are made between units. Concepts introduced early in the course are used later on. Students see cohesion from the beginning of the school year to the end. From this point of view, some skill-oriented arithmetic years at the elementary school level, some algebra courses at the secondary school level, and most proof-oriented geometry courses are quite integrated. A course that contains unrelated units is not integrated at this level of analysis, even if those units are from a variety of branches of mathematics. Therefore, it is possible to have integrated units without having an integrated course. Many curricula that are called integrated because they cover a wide variety of content are more aptly described as disintegrated at the course level because they possess no cohesion or flow.

An integrated school mathematics curriculum is of a different character than any of the smaller sizes of curriculum. It needs to—

- cover all the mathematical sciences (both pure and applied) and consumer mathematics;
- connect mathematics with the culture through its history, literature, language, and society;
- use such unifying concepts as set, proof, operation, function, transformation, algorithm, and general problem-solving strategies; and
- assume and utilize ideas from previous years in developing the curriculum of each year.

Coverage is not necessary for integration in smaller sizes of curriculum because students have more than a single year to learn mathematics and because all content cannot be covered in a single year.

Aside from foreign language curricula, the mathematics curriculum as a whole is probably the most connected of any subject curriculum. Mathematics teachers at all grade levels can reasonably count on students to have learned particular mathematical ideas and skills at previous grade levels. However, geometry, algebra, and statistics at the elementary school level tend not to be integrated. As with English or social studies or many other fields, teachers tend to ignore what was learned in previous years and begin all over as if little or nothing was learned. At the high school level, if a student can take either geometry or second-year algebra first, that curriculum has lost some of its integration because, if these subjects can be interchanged, neither is necessary for the other.

I do not know of a sustained full basal curriculum that has combined mathematics with other subjects. In 1972, I was coleader of a group at the Second International Congress on Mathematical Education in Exeter, England, dealing with links between mathematics and other school subjects. We found no ongoing basal elementary or secondary school curriculum in which mathematics and any other disciplines had been united. There were the materials of the Unified Science and Mathematics in Elementary Schools (USMES) project from the United States, but these materials were supplementary. The Open University, which had just been inaugurated in England, had such programs, but they were not at the secondary school level and were not taking the place of training in mathematics. We heard individuals report of school-level attempts to combine mathematics with other disciplines, but these attempts were always unsuccessful because not enough mathematics was able to be linked. In interdisciplinary work, pure mathematics and the study of mathematical systems tend to be ignored. The status of interdisciplinary work involving mathematics does not appear to have changed much in a generation.

## THE BEST KINDS OF INTEGRATION

THE PRECEDING examples show that the best kinds of integration are characterized by mixing ideas in a cohesive and connected manner. One might teach some probability and statistics, some algebra, some geometry, and some discrete mathematics in a given year or even in all years and have an integrated curriculum of the strands type. But if the probability and statistics or any other of the areas studied are not connected with each other, then at the level of the course, the curriculum is not integrated. This is why I argue that many standard algebra and geometry courses are quite integrated, for what is taught early in the school year is typically used throughout the year.

However, a collection of yearlong integrated courses produces a disintegrated curriculum if what is taught one year is ignored the next, and in many schools teaching has been such that ignoring the coursework of the previous year is as much the rule as it is the exception. When a student takes algebra, if the ability to solve problems using only arithmetic is ignored and the student is left to solve problems using algebraic techniques alone, then no integration of the subjects has taken place. In geometry, if the student's ability to graph and all the work of the middle school is ignored, so that—without explanation—the student is denied even the few theorems he or she knows, then there is little integration. The optimal integration takes place both within a year and from year to year.

Almost all curricula integrated by strands have a number of major weaknesses. First, because topics are not studied for long periods of time, such curricula tend to downplay the notions of mathematical systems—the fact that the truth of a statement in mathematics depends on the ability of the statement to be deduced from agreed-on principles. Probably for this reason, one does not find integrated curricula in the undergraduate and graduate courses taught to mathematics majors. Indeed, as one advances through the curriculum, mathematics courses tend to become more and more specialized.

A second weakness of most curricula integrated by strands is that geometry, particularly synthetic geometry, tends to be downplayed because it does not integrate well with algebra, and, as a result, the student seldom deals with mathematics other than in a numerical context. Given the choice, most teachers will teach algebra at the expense of geometry because they see algebra as more immediately applicable to the next mathematics their students will study. (See Flanders [1992] for a recent confirmation of this point.)

Third, if the U.S. experience at the elementary school level is any guide, by trying to cover some of every topic in every year, quite a bit of time on each topic must be spent reviewing the previous year's work to ensure that students have the necessary background. For example, if a book contains a single chapter on probability and statistics, and if this chapter is the first appearance of that topic in the curriculum, then the review presents no problem because students are not expected to have prior knowledge. But the next year, if the corresponding chapter builds on what has happened the previous year, the teacher typically needs to spend about half of a chapter's worth of class time in review. Therefore, half the time is not spent learning new material. Conversely, when a few chapters on a topic occur in a given year, then the amount of review of previous study can be nearly the same as if only one chapter appears, thereby allowing the topic to be taken much further.

Thus, the optimal integration incorporates themes and builds an integrated curriculum around those themes. This idea is in agreement with the Second International Mathematics Study and the Third International Mathematics and Science Study—both of which recommend dealing with fewer topics each year. Whether these themes are grounded in content or process, pure or applied mathematics, problems or theory, each theme must be broad enough to assimilate a variety of mathematical activities and connect with the other themes.

Even so, one must realize that integrating the curriculum is not a panacea. Integrated curricula do not necessarily lead to higher performance. For instance, students in Great Britain, which for many years has had a curriculum in which distinctions among areas of mathematics have been removed, perform no better than students in the United States do on international tests. Integrated curricula do not lead to greater retention of students in mathematics courses; the number of U.S. students who continue into precalculus mathematics or above as seniors falls in about the middle of a ranking of industrialized countries. Integrated curricula are of varying types and varying qualities. As with many educational ideas, the execution of the idea is at least as important to its value as the idea itself.

## SUMMARY

ONE OF the fundamental characteristics of mathematics is that its ideas are connected in coherent ways. Integration is a style of organizing, presenting, and doing mathematics in which openness and connections within mathematics and between mathematics and all other endeavors are encouraged. These connections may be logical, historical, or structural, and may be pure or applied. They may occur in a single problem, a lesson, a unit, a course, or the entire mathematics curriculum, or between subjects. Both traditional and recent mathematics curricula in the United States are integrated in some ways and not integrated in others. A curriculum is not integrated if a variety of unrelated activities are presented or if many strands are taught but not related to each other.

### REFERENCES

Alper, Lynne, Dan Fendel, Sherry Fraser, and Diane Resek. "Designing a High School Mathematics Curriculum for All Students." *American Journal of Education* 106 (November 1997): 148–78.

Berlin, Donna F., and Arthur L. White. "Report from the NSF/SSMA Wingspread Conference: A Network for Integrated Science and Mathematics Teaching and Learning." *School Science and Mathematics* 92 (October 1992): 340–42.

Breslich, Ernst R. *First Year Mathematics.* 4th ed. Chicago: University of Chicago Press, 1915.

______. *Second Year Mathematics.* 2nd ed. Chicago: University of Chicago Press, 1916.

______. *Third Year Mathematics.* Chicago: University of Chicago Press, 1917.

Breslich, Ernst R., Raleigh Schorling, Horace C. Wright, and Harry N. Irwin. "Course of Study in Secondary Mathematics in the University High School, University of Chicago." *School Review* 24 (November 1916): 648–74.

Bumby, Douglas, and Richard J. Klutch. *Mathematics: A Topical Approach,* Courses 1–3. Columbus, Ohio: Charles E. Merrill Publishing Co., 1978–80.

Choquet, Gustave. *L'enseignement de la géométrie.* Paris: Hermann, 1964.

Committee on the Function of Mathematics in General Education of the Commission on the Secondary School Curriculum for the Progressive Education Association. *Mathematics in General Education.* New York: D. Appleton–Century Co., 1940.

Consortium on Mathematics and Its Applications (COMAP). *Mathematics: Modeling Our World.* New York: W. H. Freeman & Co., 1998.

Core-Plus Mathematics Project. *Contemporary Mathematics in Context Sampler.* Chicago: Everyday Learning Corporation, 1999.

Coxford, Arthur F., James T. Fey, Christian R. Hirsch, Harold L. Schoen, Gail Burrill, Eric W. Hart, Ann E. Watkins, Beth Ritsema, and Mary Jo Messenger. *Contemporary Mathematics in Context—A Unified Approach,* Courses 1–3. Chicago: Everyday Learning Corporation, 1997.

Descartes, René. *Géométrie.* Translated by David Eugene Smith and Marcia L. Latham to *The Geometry of René Descartes* (1637; reprint of 1925 translation, New York: Dover Publications, 1954).

Dienes, Zoltan P., and E. W. Golding. *Geometry through Transformations,* Vol. 1–3. New York: Herder and Herder, 1967.

*Elements of Mathematics.* New York: Harper and Row, 1975.

Euler, Leonhard. *Elements of Algebra.* 1770. Translated by John Hewlett, 1840; reprint, New York: Springer-Verlag, 1984.

Fehr, Howard F., James T. Fey, and Thomas J. Hill. *Unified Mathematics,* Courses 1–4. Menlo Park, Calif.: Addison-Wesley Publishing Co., 1972.

Fiscella, Joan B., and Stacey E. Kimmel, eds. *Interdisciplinary Education: A Guide to Resources.* New York: The College Board, 1999.

Flanders, James. "Textbooks and the SIMS Test: Comparisons of Intended and Implemented Eighth-Grade Mathematics." Ph.D. diss., University of Chicago, 1992.

Hartung, Maurice L., Henry Van Engen, Lois Knowles, and Catharine Mahoney. *Seeing through Arithmetic.* Grades 1–8. Chicago: Scott Foresman & Co., 1955.

Heath, Sir Thomas. *Euclid's Elements,* Vol. 1–3. New York: Dover Publications, 1956.

Hilbert, David. *Foundations of Geometry.* Translated by E. J. Townsend. La Salle, Ill.: Open Court Publishing Co., 1902.

Hirschhorn, Daniel B., Denisse R. Thompson, Zalman Usiskin, and Sharon L. Senk. "Rethinking the First Two Years of High School Mathematics with the UCSMP." *Mathematics Teacher* 88 (November 1995): 641–47.

Hrabi, J. *Mathematics in Canadian Schools.* Ottawa: Canadian Association of Mathematics Teachers, 1967. Cited in Douglas Crawford, "Rethinking School Mathematics: 1959–Present," in *A History of Mathematics Education in the United States and Canada,* Thirty-second Yearbook of the National Council of Teachers of Mathematics (NCTM), edited by Arthur Coxford and Phillip Jones, pp. 433–34 Washington, D.C.: NCTM, 1970.

Kodaira, Kunihiko, ed. *Grade 8 Mathematics.* Translated by Hiromi Nagata, edited by George Fowler. Tokyo: Tokyo Shoseki, 1984; reprint, Chicago, Ill.: University of Chicago School Mathematics Project, 1992.

Krist, Betty. "The Gifted Math Program at SUNY at Buffalo." In *The Secondary School Mathematics Curriculum,* 1985 Yearbook of the National Council of Teachers of Mathematics (NCTM), edited by Christian R. Hirsch and Marilyn J. Zweng, pp. 177–83. Reston, Va.: NCTM, 1985.

Lappan, Glenda, James T. Fey, William M. Fitzgerald, Susan N. Friel, and Elizabeth Difanis Phillips. *Connected Mathematics, Grades 6–8.* Menlo Park, Calif.: Dale Seymour Publications, 1998.

Lee Peng Yee, Seng Teh Keng, and Keong Looi Chin. *New Syllabus D Mathematics 1. New Syllabus D Mathematics 2.* 4th ed. Singapore: ShingLee Publishers, 1997–98.

Myers, George W. "Report of the Unification of Mathematics in the University High School." *School Science and Mathematics* 11 (December 1911): 777–90.

National Committee on Mathematical Requirements. *The Reorganization of Mathematics in Secondary Education.* Washington, D.C.: Mathematical Association of America, 1923.

National Council of Supervisors of Mathematics (NCSM). "Position Paper on Basic Mathematical Skills." Distributed to members January 1977. Reprinted in *Mathematics Teacher* 71 (February 1978): 147–52.

National Council of Teachers of Mathematics (NCTM). *The Revolution in School Mathematics.* Washington, D.C.: NCTM, 1961.

______. *An Agenda for Action.* Reston, Va.: NCTM, 1980.

______. *Teaching and Learning Algorithms in School Mathematics.* 1998 Yearbook of the National Council of Teachers of Mathematics, edited by Lorna J. Morrow and Margaret J. Kenney. Reston, Va.: NCTM, 1998.

______. *Principles and Standards for School Mathematics.* Reston, Va.: NCTM, 2000.

Papy (Georges) with the assistance of Frederique Papy. *Modern Mathematics, Book I.* Translated by Frank Gorner. New York: Macmillan Co., 1968.

Paul, Fredric, and Lynn Richbart. "New York State's New Three-Year Sequence for High School Mathematics." In *The Secondary School Mathematics Curriculum,* 1985 Yearbook of the National Council of Teachers of Mathematics (NCTM), edited by Christian R. Hirsch and Marilyn J. Zweng, pp. 200–210. Reston, Va.: NCTM, 1985.

Rising, Gerald R., William T. Bailey, David A. Blaeuer, Robert C. Frascatore, and Virginia Partridge. *Unified Mathematics, Book 1.* Boston: Houghton Mifflin Co., 1981.

Rising, Gerald R., John A. Graham, William T. Bailey, Alice M. King, and Stephen I. Brown. *Unified Mathematics, Book 2.* Boston: Houghton Mifflin Co., 1981.

Rising, Gerald R., John A. Graham, John G. Balzano, Janet M. Burt, and Alice M. King. *Unified Mathematics, Book 3.* Boston: Houghton Mifflin Co., 1981.

Rutherford, E. James, and Andrew Ahlgren. *Science for All Americans.* New York: Oxford University Press, 1990.

Senk, Sharon L. "The Development of a Unified Mathematics Curriculum in the University High School: 1903–1923." Ph.D. qualifying paper, University of Chicago Department of Education, 1981.

Systematic Initiative for Montana Mathematics and Science (SIMMS). *Integrated Mathematics.* Bozeman, Mont.: SIMMS, 1996.

Usiskin, Zalman. "Lessons from the Chicago Mathematics Project." *Educational Leadership* 50 (May 1993): 14–18.

______. "On the Relations between Mathematics and Science in Schools." *The Journal of Mathematics and Science: Collaborative Explorations* 1 (fall 1997): 9–25.

# 3

# Getting Back to Our Non-Extraneous Roots

Dan Kennedy

Anyone who has chaired a high school mathematics department can probably rattle off the extant arguments against implementing an integrated mathematics program. Although the arguments take many forms, the logic is always the same. Whether on the basis of textbooks, SAT scores, college preparation, mathematics competitions, compatibility with other schools, consistency among generations, basic skills, cognitive development, or job preparedness, the background against which integrated mathematics and other reforms are measured inevitably defines mathematics education in terms of our past experience. Hence, it matters not whether the arguments against reform are any good, because they are virtually irrefutable. Logically, here is the dilemma:

- People who care about schools want students to learn more mathematics, not less.
- People who care about schools define mathematics by what is currently in the curriculum.
- By this definition, changing the curriculum results in students' learning less mathematics.
- Therefore, people who care about schools do not want to change the curriculum.

This unfortunate syllogism has led to a great deal of frustration for the many curriculum reformers who, in fact, care very deeply about schools and who have therefore adopted the simple expedient of redefining mathematics—at least for themselves. However good that idea may be, it is certainly not a fair way to win a debate. Thus, the popular arguments against curriculum reform are essentially irrefutable. School patrons will argue about whether the mathematics glass is half-empty or half-full and will agree that we ought to fill it up, but if they perceive it to be half-full of milk, they will want to fill it up with more milk. Forget about cream, and definitely forget about water. Everyone knows that what children need is milk. And so it goes.

This logic is dangerous to far more than integrated mathematics, which is really only one of many curricular reforms that seem doomed to founder on the same shoals. The core curriculum as we know it (algebra 1, geometry, algebra 2, precalculus, calculus) has become the *definition* of what needs to be done better, and as long as we stick with that definition, doing something different, no matter how well, is not likely to be seen as an improvement. This outcome is doubly ironic, first because many thoughtful mathematicians and scientists have recently put a lot of thought into what the curriculum ought to be, and second, because very little thought actually went into what the curriculum currently is.

Consider that in 1900, only about 11 percent of U.S. children between the ages of 14 and 17 attended secondary schools. By 1930, the number had risen to 51 percent, and today we have what amounts to universal public education, at least in spirit, from grades 1 through 12. In 1990, more than 110,000 schools delivered that education under the watchful eyes of roughly fifteen thousand separate school districts. Since this educational infrastructure grew up topsy-turvy in just a century, one might ask how much planning could have gone into it.

The mathematics curriculum that we have today in U.S. high schools is, for all its ancient subject matter, an artifact of the twentieth century. I used to assume that it had always been there, dating back perhaps to the ancient Babylonians, until I discovered that it had actually evolved from the decisions made by a committee of ten university academicians, known as the "Committee of Ten," in the early 1900's. Their motivation was to prepare students for college mathematics, which then meant calculus. Today, despite a century of the greatest growth in knowledge that mathematics has ever enjoyed, such preparation *still* means calculus; and so the core curriculum, defying all logic, endures in much the same form as it took when our grandparents went to school. Because the books predated the courses, a natural approach in those early days was to build the curriculum around subjects for which individual primers already existed—algebra, geometry, trigonometry, analytic geometry, and calculus. No book existed about *all* of mathematics in 1900, and hence no template for an integrated course.

Many people tend to forget that obviously no books about *all* of geometry or algebra or calculus existed either; we had books of selected topics that could serve only as introductions to those subjects. After a century of repetition, those original topics have gradually become definitive, so that many people now assume, for example, that a 900-page textbook entitled *Geometry* must surely contain everything that is important about geometry. The mere fact that nobody ever planned it that way suggests that this assumption is not likely to be true. Mathematicians should know that it is not even close.

While on the subject of circumstances that nobody ever planned, let me add that no mathematics textbook in 1900 would have contained 900 pages, either. New discoveries and the needs of the workplace have actually prompted some curricular reforms over the years, but, incredibly, they have served only to add *more* material to what was already one of the most content-laden subjects in the high school curriculum. The bloated textbooks of today consequently contain sections on applied matrix algebra, descriptive statistics, transformational geometry, linear programming, computer programming, game theory, graph theory, Boolean algebra, and optimization—topics that were unheard of forty years ago—whereas only a few of the topics from forty years ago (log tables and trig tables come to mind) have been dropped. The sad effect of adding these topics without regard to their effect on teachers and students has been chaos, with teachers' picking and choosing topics from textbooks that are far too enormous from which to teach, often omitting, or shortchanging, the very topics that the bloated textbooks were designed to preserve.

All this accumulation leads, I hope, to the conclusion that it is time to rethink the core mathematics curriculum. Defending the current curriculum on the grounds that it has always been the curriculum is not only specious but counterproductive. Defending the current curriculum on the grounds that it is a product of careful design is historically inaccurate. Defending the current curriculum on the grounds that it represents what mathematics "is" is a trivialization of mathematics. We are teaching an accident of evolution, a collection of topics that began as discrete nineteenth-century primers and that accumulated a disjointed bunch of modern topics along the way. We ought to be teaching an integrated twenty-first-century primer that is the product of careful design. Doing so would mean, however, throwing out some non-essential topics that some might consider sacred. It would entail getting back to our mathematical roots, some of which are essential and deep but some of which are, to borrow a term from algebra, extraneous.

## THE ROOTS OF OUR MATHEMATICS EDUCATION TREE

TWO ROOTS of our mathematical tree appear to be far deeper than the others. Borrowing a term from Lynn Steen (1997) and others, I call one "quantitative literacy," which consists of the knowledge that everyone ought to have to be, well, qualitatively literate. The other, I call "mathematics preparation," which consists of the knowledge that everyone ought to have to succeed in higher mathematics and perhaps someday do research in the field. Although we have no reason for thinking that mathematics education cannot draw strength from both these roots, they have become sort of a yin and yang for those in our profession who argue over such topics as calculators and two-column proofs.

However we define them, I believe both these roots to be essential and ancient. If we follow the "quantitative literacy" root all the way back, we will arrive at our primitive ancestors' bartering over tools and food. People have always studied mathematics because it is useful to the point of being necessary. If we follow the "mathematics preparation" root all the way back, we will find the eternal quest for mathematical discovery, a quest that has always begun with learners seeking out teachers. These roots are essential because we cannot have mathematics without them. Like all good roots, they provide the nourishment without which life would be impossible.

Other roots also serve important purposes. Consider the "mental development" root. Teachers in the early nineteenth century considered the development of a strong mind to be the primary goal of early schooling, and one can easily understand how they would have been drawn to mathematical computation as the ideal exercise for developing brains. But do mathematical exercises develop the brain like sit-ups develop the abs, or does rote mathematical knowledge acquired through such exercises simply enable one to do more and better mathematics? The answer to that question, which is not obvious, is important for determining the nature of the "mental development" root. We all know that today's youth can do long division, and many other manipulations, quickly and accurately on calculators. By doing so, they can avoid the tedious mental exercise that used to be necessary. That shortcut sounds like a great idea if the purpose is to find the solutions to long-division problems, but it is a terrible idea if the mental exercise is necessary, or even desirable, for developing children's brains.

I side with those who say that the "mental development" root is essential rather than extraneous. We do need to develop young brains, and rote learning is a good—perhaps indispensable—way to do so. However, I believe that that we should abandon yesterday's rote learning and replace it at every level with the rote learning that is actually necessary today. No shortage of it is to be found; in fact, far more of it exists than in the past when one considers how much more one needs to know to become an expert on anything, including mathematics. We can be smug about how well we knew our multiplication tables when we were children and wonder why students cannot memorize them today, but we should consider how many more numbers young people are apt to memorize today just to cope with our technological society: telephone numbers (often ten digits), ID numbers, PINs, security codes, TV channels, zip codes, retail prices (for comparison shopping), locker combinations, credit card numbers, and many others. We may think that our generation was better at rote procedures, but we should consider who in the family can program the VCR. Indeed, if we want to see rote procedures elevated to an art form, we can watch a fifth-grader playing a video game. We can certainly question the value of some of today's rote learning experiences, but if rote learning is valued for mental development, at least plenty of it goes on. Admittedly, nobody needs to learn video games, but let us also be prepared to admit that nobody needs to divide 563,346 by 2367 without a calculator, either. Indeed, we need to identify a lot of other things that nobody needs to know, too. The scary thing is that while we let time slip by without giving the world guidance about what sort of rote learning really is important, the world is sending our children the subtle message that all rote learning is equally valuable, regardless of the sort of knowledge gained. Having it is how one wins at Trivial Pursuit, *Jeopardy!*, and *Who Wants to Be a Millionaire.*

Because the subject has already come up, let us look at the "technology" root. Some members of our profession consider technology to be an essential part of mathematics teaching, and others think that it is a fancy distraction—possibly worthwhile, but extraneous to teaching and learning mathematics. I believe that the latter group is defining technology too narrowly. Before the electronic calculator and the computer was the slide rule. Before the slide rule was the abacus. Long before that were fingers and toes. Is finding a cosine on a calculator really all that mentally different from finding a cosine in a trig table? Is a log table considered technology if we use it to do long division without really doing long division? What about the compass, the protractor, and the straightedge? The calendar, the clock, and the astrolabe? The history of mathematics is tied inextricably to a long progression of ingenious devices that have helped scientists do their work, and part of mathematics teaching has always been to show the next generation how those devices work. In that respect, I contend that it is another essential root.

Of course, the world has changed a lot since the slide rule. Today's technology is *programmable*, giving it the capability of becoming a different tool in the hands of different mathematicians. Instruction is no longer merely a question of showing the next generation how a computer works; we really need to suggest the possibilities. Such modern mathematical breakthroughs as the proof of the four-color-map theorem and the classification of all finite simple groups were dependent on computer assistance. Entire mathematical fields of concentration, like linear programming, operations research, dynamical systems, and fractal geometry, grew up hand-in-hand with computer technology. We cannot possibly teach them all to the next generation, but the good news is that we do not have to; if we simply let them use today's technology to do the mathematics of today, they will use tomorrow's technology to do the mathematics of tomorrow.

For an extraneous root, I invite you to consider the most visible root of all: the "core curriculum" root. I will not revisit all of the points I attempted to make earlier; suffice it to say that it is extraneous. We can do whatever we want to it, and mathematics will get along just fine, just as long as we do not harm the essential roots.

Also extraneous is a root that I call the "competition" root. For some reason, mathematics has become the most competitive of the traditional subject areas, assuming that we do not classify interscholastic sports as a traditional subject area. Not only do schools spend considerable time preparing their best students for formal mathematics competitions, but many schools actually create competitions where competitions should not exist, comparing individual students, classes, schools, school districts, states, and even nations on the basis of mathematics scores taken from various assessments. The effect of all this competition on the teaching and learning of mathematics should not be underestimated. With students of lower abilities, teachers in many states spend most of their time and effort on preparation for high-stakes, minimal-competence tests. They place this emphasis not so that their students might be minimally competent—an unfortunate goal in the first place when carefully considered—but so that their students will not compare unfavorably with other students. For higher-ability students, all of us teach topics that might well fade from the curriculum were it not for the fact that our best students need to know them to answer the classical mathematics-contest questions. Such questions, deliberately designed to test cleverness and quickness, are often so far removed from reality that they are not even useful in our mainstream classroom discussions. Perhaps by concentrating on the essential roots and *giving up* some of our favorite contest topics, we can improve the teaching and learning of mathematics for all students and ultimately make the competitions better and more relevant. Meanwhile, I am going to call the "competition" root extraneous.

The final root that I will mention is more difficult to capture in a pithy name, but I call it the "culture" root. We mathematics teachers can get so dazzled by the sheer usefulness of mathematics that we forget that we are exposing our students to one of the greatest accomplishments of the human mind—a subject as rich in history as the pageants of world politics, as creative as poetry or studio art, as old as the pyramids, and as new as unlocking the genetic code. Far from extraneous, this root might be the most basic one of all. Good *cultural* and *historical* reasons can be cited to expose students to certain classical results in mathematics, regardless of whether they will need them later in life.

That addition gives the tree of mathematics (by my count) five essential roots: "quantitative literacy," "mathematics preparation," "mental development," "technology," and "culture." If we must measure proposed reforms against the background of our mathematical past, let us at least confine our attention to those essential roots rather than such extraneous roots as our unplanned, unkempt "core curriculum."

For example, let us see what happens when we consider the idea of integrated mathematics from the point of view of essential roots. We move quickly past all objections based on the facile premise that any change in the traditional, segregated, core curriculum is bad, since the "core curriculum" root is extraneous. If we concentrate on "quantitative literacy," we find ourselves drawn to multiple representations and broader applications of mathematics in richly varied contexts—both frequently cited as goals of integrated mathematics programs. If "mental development" is the goal, it is difficult to see the logic in a program that matches a child's cognitive development with subject areas (e.g., algebra, geometry) rather than with appropriate levels of abstraction—another argument frequently used on behalf of integrated mathematics. If students of higher mathematics are expected to "put it all together" mathematically, can we possibly be preparing them well by presenting it to them separately? And what about technology, which has done more to integrate mathematics in the last ten years than the two decades of philosophical discussions that preceded it? When New York state bit the bullet in the 1970s and integrated their high school mathematics courses, I naively suggested that the rest of us could simply wait ten years to see whether it caught on, not realizing at the time how extraneous roots would get in the way. Ten years later, not much had happened outside the Empire State. In a turn of events, however, the arrival of graphing calculators in 1990 has proved to be the Trojan horse of education reform, opening the door to all kinds of new ideas that suddenly seemed worthy of our consideration—integrated mathematics among them.

Although fully integrated mathematics programs are still far from common, a glance at any modern textbook series suggests that quite a lot of de facto integration is already taking place. Geometry textbooks still exist, but they are well larded with algebra, just as algebra textbooks are liberally sprinkled with applications to geometry. No textbook will sell these days without rich and varied applications in contexts that students can appreciate. Graphs and data are everywhere. Historical vignettes line the margins to give students some idea of the chronology of mathematical progress. Every textbook highlights multiple representations, and technology makes the connections easier to see than ever before. We are *all* teaching a lot more integrated mathematics than we used to, which would be a great step forward for mathematics education were it not for the unfortunate fact, noted earlier, that the textbooks now contain far too much material for teachers to teach or for students to learn. No sane person has ever have suggested that the way to integrate mathematics is to stuff twice as much material into every course, but when extraneous roots are allowed to grow alongside essential roots for long enough, insanity soon prevails anyway. We need to teach mathematics better, but we do not need bigger textbooks to do it; we need new primers.

This publication and others like it have caused more people to look beyond the core curriculum than ever before, giving us hope that a twenty-first-century primer will appear before the twenty-second century moves in. Meanwhile, our only hope as teachers seeking a little sanity is to get back to the non-extraneous roots ourselves. Therefore, I urge all of us to work with our departmental colleagues to trim courses down to a teachable set of important topics. Instead of just teaching something because it is in the book, let us ask whether it can really be included among one or more of our five essential roots:

- Does it contribute to quantitative literacy for our *current* society?
- Is it part of a *modern, broad-based* preparation for studying the mathematical sciences?
- Is it important for students' mental development—and worth knowing?
- Does it realistically incorporate the appropriate modern technology?
- Is it a significant example of mathematics as an achievement of human culture?

Those are the criteria that I plan to use when I am forced to choose between the mathematics that I will teach and the mathematics that I will, perhaps with regret, abandon. Whichever way I go, my students will miss out on something that is probably worthwhile. My only consolation will be that they will learn something essential.

Even in the days of modern computer algebra systems, we will have to find some roots for ourselves.

## BIBLIOGRAPHY

Kennedy, Dan. "Climbing Around on the Tree of Mathematics." *Mathematics Teacher* 88 (September 1995): 460–65.

Kliebard, Herbert M. "Curriculum Ferment in the 1890's" In *The Future of Education: Perspectives on National Standards in America,* edited by Nina Cobb, pp. 17–39. New York: The College Board, 1994.

National Council of Teachers of Mathematics (NCTM). *Curriculum and Evaluation Standards for School Mathematics.* Reston, Va.: NCTM, 1989.

———. *Principles and Standards for School Mathematics.* Reston, Va.: NCTM, 2000.

National Research Council, Mathematical Sciences Education Board. *Everybody Counts: A Report to the Nation on the Future of Mathematics Education.* Washington, D.C.: National Academy Press, 1989.

Steen, Lynn Arthur. "Preface: The New Literacy." In *Why Numbers Count: Quantitative Literacy for Tomorrow's America,* edited by Lynn Arthur Steen, pp. xv–xxviii. New York: College Entrance Examination Board, 1997.

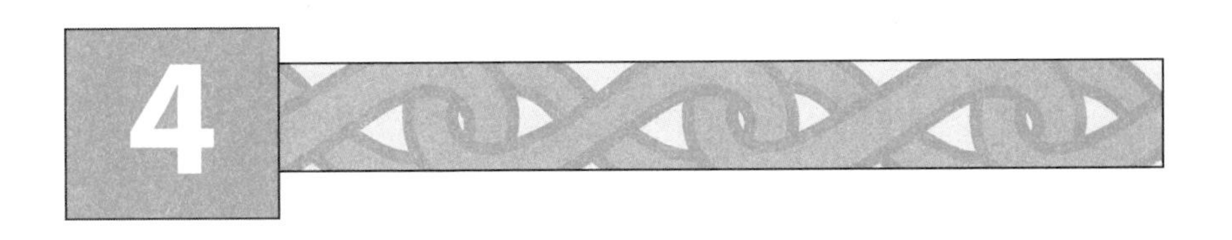


# Integrated Mathematics: From Models to Practice

Donna F. Berlin

FOLLOWING THE introduction of *Curriculum and Evaluation Standards for School Mathematics* by the National Council of Teachers of Mathematics (NCTM) in 1989, the National Science Foundation (NSF) funded thirteen mathematics curriculum projects—three at the elementary school level, five at the middle school level, and five at the high school level. The goal was to create curricular materials that reflected the content and pedagogy suggested by the *Curriculum Standards*. In particular, the NSF materials were designed to give students multiple opportunities to engage in meaningful problem solving, communication, and reasoning while connecting important mathematical concepts and procedures. These curricula offer abundant opportunities to form broader understandings within mathematics and to connect mathematical ideas with work in other disciplines and in everyday applications. Thus, these materials represent integrated mathematics curricula.

Each project produced commercial curriculum materials that are now ready for adoption by schools and districts. This chapter is divided into three sections to help teachers further their understanding of teaching with integrated materials, assist in their selection or development of integrated materials, and facilitate their implementation of integrated materials. The first section presents various models of integration. The second briefly describes the thirteen NCTM Standards–based mathematics curriculum programs, paying particular attention to features related to integrated mathematics. The chapter concludes with reflections on these programs' organization and approaches to integration.

# INTEGRATION MODELS

THE CURRENT vision of integration in teaching and learning mathematics is articulated in the Connections Standard in NCTM's *Principles and Standards for School Mathematics* (NCTM 2000, p. 64):Instructional programs from prekindergarten through grade 12 should enable all students to—

- recognize and use connections among mathematical ideas;
- understand how mathematical ideas interconnect and build on one another to produce a coherent whole;
- recognize and apply mathematics in contexts outside of mathematics.

The elaboration of the Connections Standard addresses connections within mathematics, between mathematics and other disciplines, and between mathematics and the real world.

## General Integration Models

To give a theoretical grounding for the diverse perspectives associated with integrated mathematics, it might be helpful to view integrated mathematics from the broader context of general integration models. Many of these general integration models are arranged in a continuum from single disciplines to multiple disciplines characterized by more connections (Burns and Sattes 1995; Drake 1993, 1998; Erickson 1995; Fogarty 1991a, 1991b; Jacobs 1989, 1997). For example, Burns and Sattes chart the following evolutionary stages of curriculum integration in terms of content, instruction, assessment, and classroom culture: parallel disciplines, multidisciplinary, interdisciplinary, integrated, and transdisciplinary. Drake (1998) defines the following continuum, which she suggests enjoys widespread school use: traditional, fusion, within one subject, multidisciplinary, interdisciplinary, and transdisciplinary. Erickson describes two forms of integration that progressively move students toward higher thought processes: multidisciplinary, coordinated activities curriculum, and the integrated concept/process model. Using various optical metaphors, Fogarty describes three models of integration within single disciplines (fragmented, connected, and nested); five models of integration across several disciplines (sequenced, shared, webbed, threaded, and integrated); and two models of integration within and across learners (immersed and networked). Finally, Jacobs (1997) proposes a continuum of options for content design, including discipline based, parallel disciplines, multidisciplinary, interdisciplinary units or courses, integrated day, and complete program. She suggests that these options differ by degree and nature of integration.

Clearly, numerous general models of integration exist, each with its own set of terms and definitions. Although these general models of integration are often presented in a stepwise movement toward higher degrees of integration or higher-level thinking, I do not view integration as a hierarchical continuum. Accordingly, readers are urged to review the components of each model of integration and then choose and combine those components that most closely reflect their perspectives, contexts, and purposes for choosing integrated-mathematics curricula.

## Models Integrating Mathematics with Science

Because mathematics and science are related naturally and logically in the real world, many of the NSF Standards–based mathematics curriculum programs include important content and processes needed for scientific inquiry and mathematical problem solving, as well as real-life contexts and situations that integrate mathematics with science. Furthermore, numerous models have been developed to integrate these two disciplines. As with the general models of integration, the models integrating mathematics and science most often take the form of a continuum. One of the earliest models was generated at the Cambridge Conference on Integration of Mathematics and Science Education held in 1967 (EDC 1970). Five categories of interaction between mathematics and science were identified: math for math, math for science, math and science, science for math, and science for science.

Other models have since proposed a continuum based on very similar categories. For example, Brown and Wall (1976) describe their continuum as consisting of mathematics for the sake of mathematics, mathematics for the sake of science, mathematics and science in concert, science for the sake of mathematics, and science for the sake of science. More recently, Lonning and DeFranco (1997) suggest a continuum of independent mathematics, mathematics focus, balanced mathematics and science, science focus, and independent science. Another continuum includes mathematics for the sake of mathematics, mathematics with science, mathematics and science, science with mathematics, and science for the sake of science (Huntley 1998). One final iteration of the

Cambridge model includes math for math's sake, science-driven math, mathematics and science in concert, math-driven science, and science for science's sake forth [(Roebuck and Warden 1998). Only one recent theoretical model, the Berlin-White Integrated Science and Mathematics (BWISM) Model (Berlin and White 1994, 1995, 1998), uniquely focuses on both mathematics and science in the center of the continuum. Their model identifies six aspects of integration: ways of learning, ways of knowing, content knowledge, process and thinking skills, attitudes and perceptions, and teaching strategies. This multidimensional view of the integration of science with mathematics offers a typology to describe and understand the complex nature of integration from both a content and a pedagogical position.

The presentation of the general models of integration and those specific to connecting mathematics and science is meant to encourage readers to build their own philosophies of integration. Whether readers choose to subscribe to one integration model or to combine components from various models, the resultant philosophy of integration can guide in the review, selection or development, and implementation of integrated-mathematics curricula.

The next section of this chapter briefly describes the thirteen Standards-based mathematics curriculum programs. The discussion of these programs focuses on their embodiment of integration within mathematics, integration of mathematics with other disciplines, and integration of mathematics with real-world contexts.

## STANDARDS-BASED MATHEMATICS CURRICULUM PROGRAMS

THIRTEEN STANDARDS-BASED, comprehensive mathematics curriculum programs were developed with support from the NSF. Three programs were designed for elementary school students: Everyday Mathematics (K–6); Investigations in Number, Data, and Space (K–5); and Math Trailblazers (K–5). Five programs targeted middle school learners: Connected Mathematics (6–8), Mathematics in Context (5–8), MathScape: Seeing and Thinking Mathematically (6–8), Middle Grades MATH Thematics (6–8), and Voyager Pathways to Algebra and Geometry (6–8). Five programs were developed for the high school level: Contemporary Mathematics in Context (9–12), Interactive Mathematics Program (9–12), MATH Connections (9–11), Mathematics: Modeling Our World (9–12), and Integrated Mathematics: A Modeling Approach Using Technology (9–12).

The Content and Process Standards recommended for kindergarten through grade 4 and grades 5–8 and 9–12 in *Curriculum and Evaluation*

*Standards for School Mathematics* (NCTM 1989) furnish the infrastructure for each of the mathematics curriculum programs. The programs are currently being implemented as commercial programs in schools throughout the United States. Although these mathematics programs were designed to specifically address the 1989 Standards, they remain viable and closely aligned with the vision of mathematics put forth in the more recent NCTM document, *Principles and Standards for School Mathematics* (2000).

Multiple resources—including the Web sites of each project developer and curriculum publisher, provided below—are available to describe and assist in the implementation of the thirteen Standards-based mathematics curricula. Moreover, the following three NSF implementation centers present extensive information about the mathematics curricula organized by school level:

- Elementary school: Alternatives for Rebuilding Curricula (ARC), www.comap.com/arc
- Middle school: Show-Me Center (National Center for Standards-Based Middle Grades Mathematics Curricula), www.showmecenter.missouri.edu
- High school: Curricular Options in Mathematics Programs for All Secondary Students (COMPASS), www.ithaca.edu/compass

For a basic overview of all thirteen mathematics curricula, please see *Curriculum Summaries* (June 2000) published by the K–12 Mathematics Curriculum Center at the Education Development Center, Incorporated (www.edc.org/mcc/curricula), and the database of the Eisenhower National Clearinghouse (www.enc.org).

A brief, general description of each program is followed by a discussion of features related to integrated mathematics from the perspective of NCTM's Connections Standard. The specific Connections Standard elaborations for each grade band are not reiterated here but can be found in NCTM's *Curriculum and Evaluation Standards for School Mathematics* (1989) and *Principles and Standards for School Mathematics* (2000). This chapter is not intended to provide a comprehensive review, evaluation, or endorsement of these programs but rather to highlight program features related to the Connections Standard. The descriptions of the mathematics curriculum programs were drawn from multiple sources, including program materials and information provided by the Web sites of the three NSF implementation centers, the project developers, and the publishers.

## Elementary School Projects

• **Everyday Mathematics.** As the name suggests, Everyday Mathematics uses commonplace experiences to promote the learning of arithmetic, data gathering and analysis, probability, geometry, patterns, and algebra. Designed for students in kindergarten through grade 6, the program features ten to thirteen grade-level units, each partitioned into roughly ten lessons. The units are primarily organized by mathematical topic and feature applications to real-life contexts and situations. The material is rich in opportunities for students to connect mathematics with everyday situations both in the classroom and at home. For example, mathematics is woven into such daily classroom routines as taking attendance, using the calendar, and tracking the weather. Cross-curricular connections are evident in specific units, for instance, "Mammals: An Investigation," and in theme-based projects that are interwoven throughout the school year, for instance, the "World Tour Project," both of which are explored in grade 4. Project Developer: University of Chicago School Mathematics Project Everyday Mathematics Center, everydaymath.uchicago.edu; publisher: SRA/McGraw-Hill, www.sra-4kids.com/product_info/math.

• **Investigations in Number, Data, and Space.** Designed for students in kindergarten through grade 5, this mathematics curriculum program is presented through inclusive teachers' books, one for each of the six to eleven units at each grade level. These units focus on one or more mathematical topics that are explored through a series of activity-based investigations designed to develop students' understanding of number, data, geometry or space, and the mathematics of change. Because the students work in-depth on a number of connected investigations at each grade level, they have opportunities to link conceptual and procedural knowledge, various representations of concepts and procedures, and different mathematical topics. For example, second-grade students connect mathematical ideas in geometry and fractions by investigating the structure of two- and three-dimensional shapes in the "Shapes, Halves, and Symmetry" unit. Project Developer: TERC, www.terc.edu/investigations; publisher: Scott Foresman, www.scottforesman.com.

• **Math Trailblazers.** This mathematics curriculum for students in kindergarten through grade 5 offers connections within areas of mathematics as well as strong cross-curricular connections with science and language arts. Using real-world, multisolution problems presented in sixteen to twenty grade-level units containing multiple lessons, students explore concepts that are fundamental to both mathematics and science, including variable, length, area, volume, mass, and time. The units are primarily organized by mathematical topic. Language arts skills are developed through connected reading, writing, and communication activities. Scientific investigations embedded in many units use the four steps of the Teaching Integrated Mathematics and Science (TIMS) Laboratory Method: draw a picture, gather and organize data, graph the data, and analyze the results. For example, in unit 5, "Area of Different Shapes," third-grade students use the concept of area and the TIMS Laboratory Method to decide which brand of paper towel is more absorbent in "The Better 'Picker Upper' experiment." Project Developer: Teaching Integrated Mathematics and Science (TIMS) project at the Institute for Mathematics and Science Education, University of Illinois at Chicago, www.math.uic.edu/IMSE; publisher: Kendall/Hunt Publishing Company, Elementary-High School Division,www.kendallhunt.com/elhi.

## Middle School Projects

• **Connected Mathematics.** As the title suggests, a main goal of this program for students in grades 6–8 is to develop significant and meaningful connections related to mathematics. Throughout the materials, connections are developed among the core ideas of mathematics, between mathematics and other school subjects, among classroom activities and the interests of middle school students, and between mathematics and its applications outside the classroom. The materials, consisting of eight units at each grade level, address five content strands: number, geometry, measurement, algebra, and statistics and probability. In each unit, students explore a series of four to seven connected investigations to develop a major concept or cluster of concepts in mathematics. A set of one to five problems involving real applications, whimsical settings, or mathematical problem situations is embedded within each investigation. An example of a problem is "Pricing Pizza" (problem 7.1), which presents a real-world context that is interesting and relevant to middle school learners. This problem is one of five problems in "Going Around in Circles Investigation" (investigation 7), designed to develop connections among a circle's diameter, radius, area, and circumference in the "Covering and Surrounding Unit" (grade 6). Project Developer: Connected Mathematics Project at Michigan State University, www.math.msu.edu/cmp; publisher: Prentice Hall, www.phschool.com/math/cmp.

• **Mathematics in Context.** Mathematics in Context includes forty instructional units for students in grades 5 through 8; ten developed for each grade level. As the name suggests, a guiding feature of this program is the extensive use of realistic contexts. Based on the theory of Realistic Mathematics Education, a theoretical approach developed by the Freudenthal Institute, the underlying philosophy of this program is that mathematics should make sense to students and that they should use mathematics to make sense of the world around them. Consequently, each instructional unit features four to eight sections containing problem scenarios and related problems that furnish an engaging, real-life context in which to learn and apply mathematics. Concepts from the four content strands of number, algebra, geometry, and statistics are explored in-depth and connected across the instructional units. For example, "Dry and Wet Numbers" (grade 5 algebra strand) introduces the concepts of positive and negative numbers through investigations involving ships moving through canal locks, movements on a grid map, and surveying land, whereas the "Cereal Numbers" unit (grade 7 number strand) presents a context in which students revisit the concepts of number (including whole numbers, percents, decimals, and fractions) and the geometric concepts of volume and surface area as they solve problems about corn production, marketing, packaging, economics, graphics, and nutrition. Project Developers: Wisconsin Center for Education Research at the University of Wisconsin—Madison, www.wcer.wisc.edu, and the Freudenthal Institute at the University of Utrecht, Netherlands, www.fi.uu.nl; publisher: Encyclopaedia Britannica, Incorporated, www.ebmic.com/mic/common/home.asp.

• **MathScape: Seeing and Thinking Mathematically.** The underlying philosophy of this mathematics curriculum designed for students in grades 6–8 is that mathematics is a human experience. Students experience mathematics through real-world investigations that involve designing, creating, predicting, investigating, analyzing, evaluating, communicating, and other human endeavors. Each of the seven grade-level units consists of twelve lessons and focuses on one or more of the four broad content areas of number and operations, geometry and measurement, algebraic thinking, and data analysis. The main component of each lesson is a hands-on investigation. For example, in the "Mathematics of Motion" unit (grade 8), students measure motions and investigate the concepts of relative distance, absolute distance, and average speed. Connections are made between the content areas of algebraic thinking and data analysis as students represent motion as rates, diagrams, maps, written descriptions, equations, and distance-time graphs. Project Developer: Seeing and Thinking Mathematically Project at the Education Development Center in collaboration with Creative Publications, www2.edc.org/MathscapeSTM; publisher: Creative Publications, www.glencoe.com/sec/math.

• **Middle Grades Math Thematics.** As its name suggests, this mathematics curriculum for students in grades 6–8 is structured around eight thematic modules for each grade level. The modules are designed to help students see mathematics in their daily lives, connect mathematics with other disciplines, and connect a variety of mathematical ideas with one another. Mathematical concepts associated with ten content strands—algebra, coordinate geometry, discrete mathematics, geometry, measurement, measurement-geometry, number, probability, problem solving, and statistics—are developed in the context of relevant and meaningful everyday applications. In "Statistical Safari" (module 3, grade 6), the theme of animals furnishes the context in which students explore sets and metric measurement; fractions and percents; bar graphs and line plots; mean, median, and mode; estimation; dividing by decimals; mental math; and stem-and-leaf plots. Project Developer: The Sixth Through Eight Mathematics (STEM) Project at the University of Montana—Missoula, lennes.math.umt.edu/~stem; publisher: McDougal Littell, www.classzone.com/start/math_thematics.cfm.

• **Pathways to Algebra and Geometry.** Designed for students in grades 6-8, this mathematics curriculum is divided into two courses to be completed before entering an algebra course. Course 1 is designed for students in grades 6 or 7 and course 2 is designed for students in grades 7 or 8. Each course is organized around challenging, complex, significant, real-world situations. Two types of units are featured: (1) core applications projects in which mathematical concepts and skills are developed in extended role-plays of real-life situations, and (2) side trips that provide extensions of the mathematics in the projects or investigations of other mathematical concepts. Although all thirteen Standards for grades 5–8 (NCTM, 1989) are addressed in this program, the topics of proportional reasoning and of algebra and functions are central and are developed and revisited through many of the units. An example of a core applications project is "The Antarctica Project" (grades 7 or 8), in which students learn about scale, proportion, and algebra by role-playing architects designing a research station for scientists living and working in the frozen south. Project Developer: Middle School Math through Applications Program (MMAP), mmap.wested.org; publisher information: West Ed, mmap.wested.org/pathways.

## Secondary School Projects

• **Contemporary Mathematics in Context.** Instead of offering course sequences designed for the traditional tracks of "college-prep," "tech-prep," or "work-prep," Contemporary Mathematics in Context presents a three-year core curriculum for all students in grades 9, 10, and 11. A flexible fourth-year course is designed to prepare students for college mathematics. Each yearlong course in the core curriculum contains seven units and a two-week, thematic, project-oriented "capstone" that applies mathematics to a real-world situation, such as "Forests, the Environment, and Mathematics" in course 2. The fourth-year course contains ten units. The courses' integrated mathematics includes four strands: algebra and functions, geometry and trigonometry, statistics and probability, and discrete mathematics. These strands are further connected through such mathematical ideas as function, such mathematical habits of mind as providing convincing arguments, and such mathematical themes as representation. The program is organized primarily around big mathematical ideas embedded in rich, applied problem contexts. A central component of this program is investigations, both inside and outside class, that are rooted in real-life contexts and situations. Major themes evidenced throughout the program include mathematical modeling and mathematics as sense-making. Project Developer: Core-Plus Mathematics Project (CPMP) at Western Michigan University, www.wmich.edu/cpmp; publisher: Glencoe/McGraw-Hill, www.glencoe.com/sec/math.

• **Interactive Mathematics Program (IMP).** This four-year mathematics program is an integrated curriculum both for students who are college-bound and for those preparing to enter the workforce. In contrast with the traditional algebra 1, geometry, algebra 2/trigonometry, precalculus sequence of mathematics courses, IMP uses a problem-centered approach to draw simultaneously from the traditional areas of mathematics and such additional areas as statistics, probability, curve fitting, and matrix algebra. Each year, students complete five units involving different complex core problems and various mathematical concepts and skills. Some of the problems are realistic and practical, for instance, analyzing polling results (year 4, unit 5, "The Pollster's Dilemma"); others provide a cross-disciplinary context, for example, determining the best design for a honeycomb (year 2, unit 3, "Do Bees Build It Best?"); and finally, others are more whimsical, for instance, determining the release of a diver to create a "splash" instead of a "splat" (year 4, unit 1, "High Dive"). Project Developer: Interactive Mathematics Program, www.mathimp.org; publisher: Key Curriculum Press, www.keypress.com.

• **Math Connections.** This three-year core curriculum blends the topics of algebra, geometry, probability, statistics, trigonometry, and discrete mathematics through a unified, interconnected, concept-driven approach. Although students solve real-world problems in real-world settings, the program is based on general mathematical themes rather than problem-centered themes. The program features two books at each grade level, and the general mathematical themes are "Data, Numbers, and Patterns" (grade 9), "Shapes in Space" (grade 10), and "Mathematical Models" (grade 11). In addition to the robust connections made between the different mathematics topics, connections are made with real people and everyday situations; with the world of science and technology; and with other subjects, such as history, geography, language, and art. For example, in book 2a, students in grade 10 explore the sine function as they investigate the optimum height for windmills used to generate electricity in California. Project Developer: Math Connections: A Secondary Mathematics Core Curriculum at the Connecticut Business and Industry Association (CBIA) Education Foundation,www.mathconnections.com; publisher: It's About Time, Incorporated, www.its-about-time.com.

• **Mathematics: Modeling Our World.** This integrated mathematics curriculum program for students in grades 9–12 is organized into four courses containing seven or eight units, each further subdivided into four to seven lessons. The units are arranged by context and application rather than by mathematical topic. For each unit, major contextual themes, major mathematical themes, related disciplines, and a scope-and-sequence chart of mathematics concepts are provided. This program is rooted in real-world situations and problems and emphasizes the use of mathematical modeling throughout. For example, ninth-grade students develop models to predict human stature on the basis of forearm length or stride length (course 1, unit 4: "Prediction"), and tenth-grade students use geometric concepts to develop packaging models for holding soft-drink cans (course 2, unit 4: "The Right Stuff"). Project Developer: The Consortium for Mathematics and Its Applications, www.comap.com; publisher: Bedford, Freeman and Worth Publishing Group, www.whfreeman.com/highschool/mathematics.

• **Integrated Mathematics: A Modeling Approach Using Technology** This curriculum is designed to replace all mathematics courses for students in grades 9–12, except for Advanced Placement courses. Each of its six yearlong levels includes thirteen to sixteen thematic modules. Levels 1 (grade 9) and 2 (grade 10) are core curricula recommended for all students. Levels 3 (grade 11) and 5 (grade 12)—which focus more on applications from business and the social sciences—are intended for students who plan to pursue nonmathematical and scientific fields. Levels 4 (grade 11) and 6 (grade 12) are intended for students who plan to pursue a career in mathematics and science and are recommended for students who desire calculus as a next course. Both traditional topics (such as algebra, analysis, data analysis, geometry, matrices, statistics, trigonometry, and probability) and less-traditional topics (such as graph theory, game theory, and chaos theory) are integrated throughout the six levels. The materials emphasize the relationships among topics within mathematics and between mathematics and other disciplines through an applied, real-world context.

Two additional integral features of the program are evoked by the subtitle of the materials, "A Modeling Approach Using Technology." The following two examples are reflective of the goals of the Integrated Mathematics program: (1) Students in level 2 (grade 10) use a geometry utility to investigate geometric properties of a circle as they reconstruct a Native American medicine wheel in the "Traditional Design Module," and (2) students in level 4 (grade 11) use a graphing utility to investigate the functions required to model cyclic events, such as the motion of a mass on a spring in the "Can It! Module." Project Developer and Publisher: SIMMS Integrated Mathematics (IM) project at the University of Montana—Bozeman in association with Pearson Custom Publishing, www.montana.edu/wwwsimms.

## FINAL REFLECTIONS

MOST OF the thirteen Standards-based, comprehensive mathematics programs are organized by a series of grade-level, yearlong or course-long units that can be further partitioned into connected lessons, investigations, or problems. One high school program is divided into two yearlong core levels to be followed by a choice of levels dependent on the student's career aspirations. The units are most often structured around one or more broad mathematical content areas, strands, or topics. The mathematics includes not only the traditional content outlined in the NCTM's *Standards* document (1989) but also such nontraditional topics as discrete mathematics and chaos theory. In contrast, thematic modules or applications projects are the organizational focus for two of the middle school programs. Three high school programs use a similar contextual structure. Regardless of the organizational schema, these curriculum programs are mathematics-driven and use themes, problems, or investigations as the contexts in which to develop and apply the mathematics. The programs focus on the mathematics content and processes outlined in the NCTM's 1989 *Standards* and promote the connectedness of mathematical topics and ideas.

The thirteen curriculum programs emphasize the application of mathematics to real-world situations and experiences though activity-based investigations that involve multiple strategies and solutions. The investigations most often feature real-world or mathematical-problem situations, although a few programs employ whimsical settings. All the problems are meant to be engaging, interesting, appealing, and relevant to students in their targeted grade level. This focus is particularly true at the elementary and middle school levels, where the problems are often embedded within everyday, commonplace applications and routine class and home situations. In the upper grades, the problems tend to be more complex and may involve significant, controversial issues and extended role-playing scenarios. Mathematical modeling is an essential feature of the secondary-level programs.

Cross-curricular connections are evident in all the programs in that they seek to link mathematics with content and situations in other disciplines. The connections with other disciplines are embedded most often within the context of real-world applications. As such, the cross-curricular connections tend to be contextual rather than conceptual. At the elementary school level, the mathematics programs often forge connections with science and language arts. Although many applications are drawn from the field of science, many other disciplines furnish a context in which to explore mathematical topics and ideas. This pattern is especially true at the secondary school level, where connections are made with such fields as architecture, English, history, political science, and the social sciences, to name a few.

In closing, all thirteen Standards-based programs reflect an integrated-mathematics approach. They make connections among mathematical topics and ideas and connect mathematics with other disciplines and with the real world. Whether schools or districts adopt one of these mathematics programs or develop their own integrated-mathematics activities, decisions regarding the nature and degree of integration need to be made. The theoretical models of curriculum integration presented earlier in this chapter can assist the reader in shaping and in guiding curricular design and implementation.

## ACKNOWLEDGEMENT

The author wishes to sincerely thank Anita Bowman for her thoughtful reviews of an earlier version of this chapter.

## REFERENCES

Berlin, Donna F., and Arthur L. White. "The Berlin-White Integrated Science and Mathematics Model." *School Science and Mathematics* 94 (January 1994): 2–4.

Berlin, Donna F., and Arthur L. White. "Connecting School Science and Mathematics." In *Connecting Mathematics Across the Curriculum,* 1995 Yearbook of the National Council of Teachers of Mathematics (NCTM), edited by Peggy A. House and Arthur F. Coxford, pp. 22–33. Reston, Va.: NCTM, 1995.

Berlin, Donna F., and Arthur L. White. "Integrated Science and Mathematics Education: The Evolution of a Theoretical Model and Implications for Instruction and Assessment." In *International Handbook of Science Education,* edited by Barry J. Fraser and Kenneth Tobin, pp. 499–512. Dordrecht, Netherlands: Kluwer Academic Publishers, 1998.

Brown, William R., and Curtiss E. Wall. "A Look at the Integration of Science and Mathematics in the Elementary School—1976." *School Science and Mathematics* 76 (November 1976): 551–62.

Burns, Rebecca Crawford, and Beth D. Sattes. *Dissolving the Boundaries: Planning for Curriculum Integration in Middle and Secondary Schools and Facilitators' Guide.* Charleston, W.Va.: Appalachia Educational Lab, 1995.

Drake, Susan M. *Planning for Integrated Curriculum: The Call to Adventure.* Alexandria, Va.: Association for Supervision and Curriculum Development, 1993.

————. *Creating Integrated Curriculum: Proven Ways to Increase Student Learning.* Thousand Oaks, Calif.: Corwin Press, 1998.

Education Development Center (EDC). *Final Report of Cambridge Conference on School Mathematics, January 1962–August 1970.* Cambridge, Mass.: EDC,1970. (ERIC Document Reproduction Service no. ED 188 904)

Erickson, H. Lynn. *Stirring the Head, Heart, and Soul: Redefining Curriculum and Instruction.* Thousand Oaks, Calif.: Corwin Press, 1995.

Fogarty, Robin. *The Mindful School: How to Integrate the Curricula.* Palatine, Ill.: Skylight Publishing, 1991.

————. "Ten Ways to Integrate Curriculum." *Educational Leadership* 49 (October 1991): 61–65.

Huntley, Mary Ann. "Design and Implementation of a Framework for Defining Integrated Mathematics and Science Education." *School Science and Mathematics* 98 (October 1998): 320–27.

Jacobs, Heidi Hayes. *Interdisciplinary Curriculum: Design and Implementation.* Alexandria, Va.: Association for Supervision and Curriculum Development, 1989.

Jacobs, Heidi Hayes. *Mapping the Big Picture: Integrating Curriculum and Assessment K–12.* Alexandria, Va.: Association for Supervision and Curriculum Development, 1997.

Lonning, Robert A., and Thomas C. DeFranco. "Integration of Science and Mathematics: A Theoretical Model." *School Science and Mathematics* 97 (April 1997): 212–15.

National Council of Teachers of Mathematics (NCTM). *Curriculum and Evaluation Standards for School Mathematics.* Reston, Va.: NCTM, 1989.

————. *Principles and Standards for School Mathematics.* Reston, Va.: NCTM, 2000.

Roebuck, Kay I., and Melissa A. Warden. "Searching for the Center of the Mathematics-Science Continuum." *Science and Mathematics* 98 (October 1998): 328–33.

5

# Making Connections within, among, and between Unifying Ideas in Mathematics

Lisa Clement
Judith Sowder

THE MATHEMATICAL terrain for students in kindergarten through grade 12 is laden with connections among unifying ideas, but students do not often have opportunities to integrate these connections with their learning. This chapter illustrates, through examples, how teachers might discuss unifying ideas with students in ways that explicitly connect those ideas with other *unifying* ideas. The term unifying captures the sense that mathematical ideas are inherently interconnected and interwoven.

Unifying ideas can focus on either mathematics content or mathematical processes. For example, measurement is a unifying idea because it is connected with other ideas and weaves them together—*numbers* in contexts are measures, measures can be *represented* in a variety of ways, *data analysis* requires the study of measures, measurement is a process used for *quantifying* characteristics of our experiences, and so on. Representing mathematical ideas is a unifying idea that may be thought of as a process. For example, *data* can be represented in different forms to be analyzed, *place-value* ideas emerge from students' use and understanding of different representations of the base-ten system, understanding *functions* resides in students' abilities to connect graphical, tabular, verbal, and written representations, and so on. The greater the number and strength of the connections that students create among unifying ideas, the better they will understand them (Hiebert and Carpenter 1992). The process of seeking connections is a generative one—the greater the number of opportunities that students have to make connections, the more likely they will be to search for future connections themselves.

We have taken two unifying ideas—quantity and measurement—and described a problematic issue within each one. We then explore reasons why the issue is important for children to grapple with, examine the issue more closely in an instructional example, and highlight the ways in which this issue is not just part of one unifying idea but is connected with other unifying ideas.

## QUANTITY

FEY (1990) described the need to reconceptualize and expand the notion of *quantity* and how it fits into instruction for students in kindergarten through grade 12. He cited technological advances and a greater knowledge base about how children think about mathematics, including the difficulties they often have, as two reasons for the need to change. For example, often when students tackle a new word problem, many of them focus only on the numbers given in the situation to decide on an operation to perform without paying much attention to what the numbers are measuring or counting. Sowder (1988) found that some upper-elementary-level students would look at the relative size of the numbers in a word problem to determine an operation to perform—if the numbers were close together, they would add or multiply; if the numbers were far apart, they would divide or subtract. Other students would search for key words to solve a word problem. Consider, for example, the following situation:

> Two people who have been on diets are talking:
>
> *Dieter A:* "I lost 1/8 of my weight. I lost 19 pounds."
>
> *Dieter B:* "I lost 1/6 of my weight, and now you weigh 2 pounds less than I do."
>
> What was dieter B's original weight?

Although many students are able to figure out dieter A's weight before the diet, some choose to multiply 1/8 and 19 because the key word *of* signifies multiplication to them. Other students choose to subtract 2 pounds from dieter A's final weight to find dieter B's final weight because the words *less than* obligate them to subtract. After completing these computations, students are left with a number, yet most are unsure what the number represents within the context of the problem. Rather than try to make sense of the mathematical situation, students attempt to take the collection of numbers and find the correct operations to perform on them. Many students have learned to survive in mathematics courses by memorizing rules and using key words to get answers to word problems. However, as soon as problems involve more than one operation, these strategies fail them.

Thompson (1994) introduced the notion of *quantitative analysis* to address the fact that many students in mathematics classes use immature, non-sense-making strategies to solve such problems as those we have described. These strategies inhibit not only their ability to solve problems well but also their ability to understand the nature of numbers and operations. Thompson made an important distinction between a quantity and a value: a *quantity* is anything that can be measured or counted; the *value* of a quantity is its measure or the number of items that are counted and involves a number and usually a unit. In the Dieter's Problem, some examples of quantities include dieter A's weight before the diet; the number of pounds that dieter A lost; the fraction of weight that dieter B lost; the difference in weights between dieter A's after-diet weight and dieter B's after-diet weight; the ratio of dieter B's after-diet weight to her before-diet weight, and so on. Notice that quantities can be created by combining two other quantities additively or multiplicatively and never involve numerical values. Examples of values would include, correspondingly, 152 pounds, 19 pounds, 1/6 of dieter B's before-diet weight, 2 pounds, 5/6, and so forth. Using this language, then, one would say that many

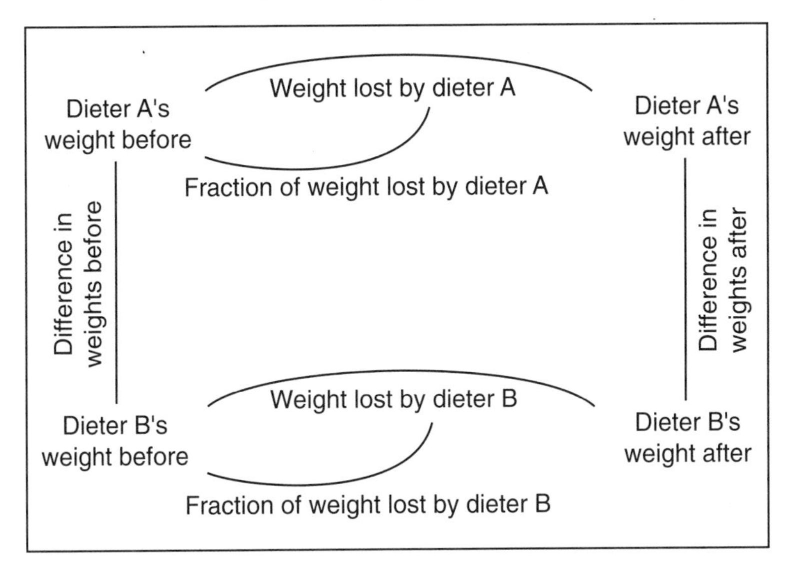


Fig. 5.1. One representation of the quantitative structure of the Dieters' Problem

students focus too closely on the *values* involved in a situation instead of reasoning about the *quantities* to which these values refer. Thompson describes *quantitative analysis* as the process of coming to understand the quantities and relationships among quantities in a mathematical situation. One way to represent relationships among quantities is shown in figure 5.1.

The ability to analyze situations quantitatively is a unifying idea for students in kindergarten through grade 12 because students who can reason quantitatively know why they are choosing the appropriate operations when solving a problem and because they must use number strategies to develop correct responses. Ample evidence exists showing (a) that even children in kindergarten through grade 2 can and do reason quantitatively when given opportunities to do so and (b) that the emphasis on reasoning quantitatively rather than searching for key words to determine the appropriate operation helps students make sense of situations (Carpenter et al. 1989).

## AN INSTRUCTIONAL EXAMPLE FOR PREKINDERGARTEN THROUGH GRADE 2

IN THE video *Making Meaning of Operations* (Teaching to the Big Ideas 1997), a teacher posed the problem "How many legs do three elephants have?" The children in the class solved the problem in various ways (the class in the videotape is a third-grade class, but the problem is appropriate for lower grades). One girl drew three circles with four line segments to represent the legs. She then wrote, "4 times 3 equals 12." Although the student successfully solved this problem, the teacher continued to query the student about the quantities involved in the situation rather than focus solely on the numbers. The teacher asked, "What does the '4' stand for in your number sentence? What does the '3' stand for in your number sentence?" This teacher recognized that the important aspect of this problem was understanding what *quantities* the *values* represented, because she knew that from this understanding, the operations would emerge in a meaningful way for students who were not yet able to solve the problem. In prekindergarten through grade 2, when teachers focus on the thinking of individual children and create situations for them to make sense of, these students are better problem solvers than, and know as many number facts and operations as, children in classrooms where teachers place less emphasis on problem solving (Fennema et al. 1996).

## CONNECTIONS WITH OTHER UNIFYING IDEAS

IN THE example from the video, students were given opportunities to connect their understanding of the *quantities* in the situation with their understanding of *operations* and *numbers.* When one looks at connections with the other unifying ideas, one might notice that students were also *problem solving* and *communicating* about their thinking. They were given opportunities to *represent* their thinking in multiple ways, as the girl did in the example. She drew a model, or picture, of the situation and wrote a correct number sentence. Notice that the task per se — finding the number of legs that three elephants have— did not lend itself to these connections among unifying ideas; rather, the ways in which the teacher introduced the problem and asked questions to explore specific content areas and processes did so. For example, the teacher gave the students the opportunity to solve the problem in ways that made sense to them (*problem solving*) using whatever representation that made most sense to them (*representation*), questioned the students about the quantities that the numbers represented (*quantity*), asked why the operations they chose made sense (*quantity*), and had the students verbally articulate their responses (*communication*). She could have pushed their thinking to become more algebraic in nature if she had asked such a question as "Suppose I call the zookeeper tomorrow to find out the number of elephants at the zoo. Is there a way that I can figure out the number of legs that those elephants have altogether?" Because students would not know the exact number of elephants, they would need to reason through this problem by extending their quantitative reasoning to involve *algebraic* reasoning without the formal notation of algebra. For example, a child might respond, "However many elephants there are, take that number and multiply by 4, which is the number of legs on each elephant." The interactions among the teacher, the task, and the students allowed students the opportunities to forge connections among unifying ideas. This interaction was not merely about solving one problem but also about allowing students to think about the integration of the unifying ideas that the task affords. Table 5.1 illustrates the task, the unifying idea, and other unifying ideas that may be integrated with it.

TABLE 5.1. QUANTITY AND OTHER UNIFYING IDEAS

| Grade Band | Major Unifying Idea | Problem | Connections with Other Unifying Ideas |
|---|---|---|---|
| Pre-K–2 | Quantity | How many legs do three elephants have? | •Number and operations<br>•Problem solving<br>•Communication<br>•Representation<br>•Algebra |

# MEASUREMENT

THE UNIFYING idea of measurement, that is, representing a characteristic by a number, has within it many ideas that are often left implicit in its teaching. In a typical measurement lesson, students are asked to find the length, area, or volume of geometric objects and are given standard units for measuring and procedures for doing so (Fey 1990). But some other important ideas in measurement include the approximate nature of real measurements due to limitations of the measurement instrument and human error; the initial arbitrariness of the unit chosen as the standard for measuring; the ability of students to make decisions about the most appropriate unit to use (e.g., centimeters versus kilometers); why and under what circumstances some measures are more meaningful than others (e.g., differences versus ratios); and how to produce a valid way to measure a characteristic, such as intelligence or quarterback quality (Moore 1990).

Many of these ideas are often overlooked when measurement is taught, but their inclusion can give students insights into, and greater understandings of, mathematics and the world in which they live. For example, certain characteristics of objects, people, countries, and the like are measured every day. These characteristics include measurements of economic indicators (e.g., unemployment rate, families' incomes, savings rates of Americans), a person's health (e.g., blood pressure, lung capacity, weight-to-height comparisons, temperature), or an athlete's performance (e.g., earned-run average or win-loss record of baseball pitchers, gymnasts' scores on such events as the floor exercise, swimmers' times in races). Understanding how these measurements are made (e.g., asking, "How is the inflation rate measured?"), determining the most appropriate measure for a given circumstance (e.g., asking, "Which measure—earned-run average, win-loss record, or something else—best describes a baseball pitcher's performance during a season?"), and developing new ways to measure quantities (e.g., asking, "How would you measure a city's livability?") are all part of understanding aspects of our own lives. Also, by grappling with many of these measurement issues, students have opportunities to develop the ability to reason proportionally—an area in which many middle and high school students traditionally have considerable weaknesses.

## AN INSTRUCTIONAL EXAMPLE FOR GRADES 6–8

THE FOLLOWING task was designed to elicit and enhance students' thinking about determining the most appropriate measure for a given situation. It is different from most measurement tasks in that students are asked to determine the best measure and explain the reasoning underlying their decision rather than told the type of measurement process they should use and asked to perform calculations. This task, when appropriately taught and learned, compels students to consider those occasions when ratio is the most appropriate measure to use to quantify an object's characteristic, in this example, squareness.

> Squareness problem: A new housing subdivision offers rectangular lots of three different sizes:
>
> 1. 75 feet by 114 feet
> 2. 455 feet by 508 feet
> 3. 185 feet by 245 feet
>
> If you were to view these lots from above, which would appear most square? Which would appear least square? Explain your answers. In what attribute or characteristic of the lots are we interested? In what ways can this amount be quantified? (Simon and Blume 1994)

For students to solve this task, they must determine the *squareness* of a rectangle—one quantity for which a ratio is the appropriate measure. Task creators Simon and Blume believe that the result of the teachers' always stating which measuring process to use might have serious and perhaps negative implications for students. That is, if told, students might come to see ratio as that which the teacher has decided to take a measure of rather than that which has an appropriate use as a measure, of squareness, for example. We expect that by considering such problems as the squareness problem, students will begin to reflect on the reasoning behind the operations that they perform before they perform them.

When students initially solve this problem, they most often come up with one of two methods. The first is the difference method: Students find the difference between the length and the height of the rectangle. The closer the difference is to zero, the more nearly square the rectangle is. The second is the ratio method: Students find the ratio between the length and the height of the rectangle. The closer the ratio is to 1:1, the more nearly square the rectangle is.

Students can reason about their respective methods, focusing on the quantities in the problem. For example, students who find the difference between length and height reason that the rectangle that is most nearly square would have a difference in dimensions closer to zero than the other rectangles. Students who find the ratio of the length and height reason that the ratio of these dimensions is 1:1 when the rectangle is square, so the ratio of the dimensions of the most square rectangle will be closest to 1:1.

On the surface, both methods seem plausible to students, but the students find that the two methods yield different rectangles as being most square. They then have to reason about a process for deciding which method is most appropriate for the situation. To help students with this process, a teacher might pose the question "What are some things we could do mathematically to determine which method is better?" Students must think about how they would choose examples to test the two methods. The teacher might next ask, "Does it matter which examples we test, or are some examples better than others?" Choosing examples reasonably to test methods or conjectures is often difficult for students and something that they are not often asked to do. For example, the idea of selecting extreme examples and holding one variable, such as the difference in the dimensions, constant (e.g., comparing the squareness of a 1-unit-by-10-unit rectangle with that of a 100-unit-by-109-unit rectangle) may take a long time for students to value as being essential to helping determine which process is better, but it is an important part of understanding the field of mathematics and what mathematicians do. Students can then make tables with a column for each example; measures using the difference method; measures using the ratio method; and determining most square to least square, the latter as determined both by the methods and by using such visual representations as graph paper to make sketches of the rectangles. The activity fosters students' insights into why ratio is the better method. Although a formal, deductive proof is not involved, tasks like this are appropriate for students in grades 6–8 and may lend insights to rigorous mathematical proof in later years.

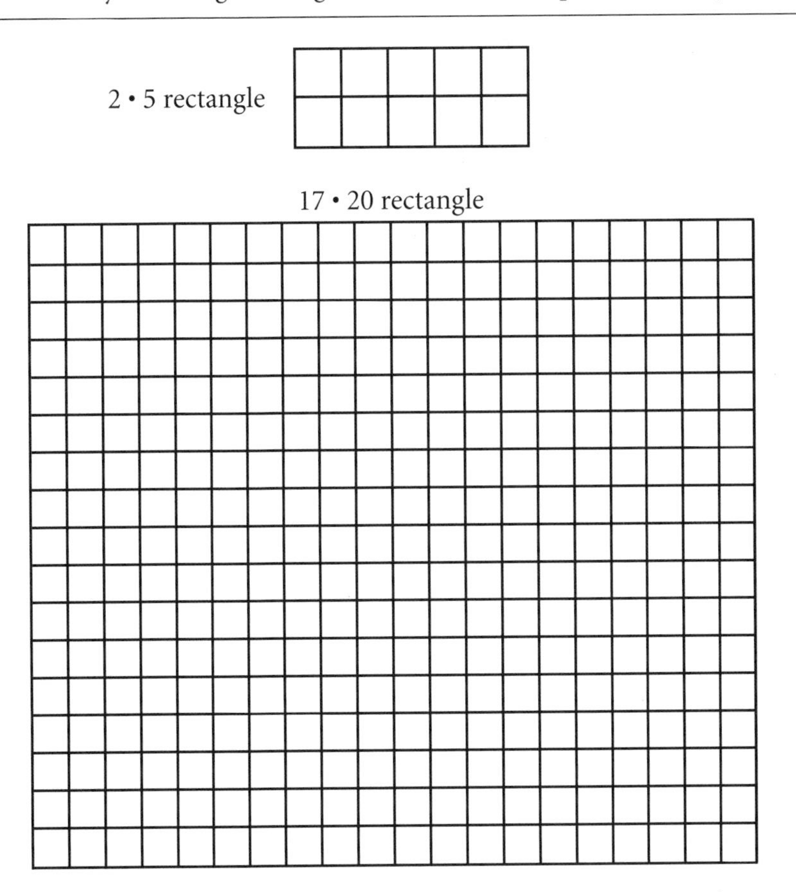


Selecting examples to hold one measure of interest constant is one feature of mathemematizing this problem. Although the difference in the dimensions of both rectangles is 3 ($2 \times 5$ and $17 \times 20$), students can visually examine the plots to see that the second plot is clearly more square than the first plot, and thus that the difference is not a good measure of squareness. The ratio of the dimensions, which are .4 and .85, respectively, is a better measure of squareness than the difference of the dimensions.

Fig. 5.2. The difference method and the ratio method

## CONNECTIONS WITH OTHER UNIFYING IDEAS

THE SQUARENESS Problem affords many connections between and among unifying ideas and helps students integrate these ideas. Students grapple with selecting the most appropriate measure—a major idea in *measure-*

*ment* that requires them to reflect on relevant quantities related to rectangles and squares (*quantity*). Because students are not told which method to use, they must *problem solve.* They may use tables, words, and sketches (*multiple representations*); identify characteristics of rectangles and squares and create drawings of rectangles (*geometry*); choose examples in a reasonable way to test the two methods (*reasoning*); and interpret and analyze the data collected from the examples to determine the better measure (*data analysis*) to appropriately accomplish this task. Table 5.2 lists the task, the unifying idea, and other unifying ideas that can be integrated with it.

TABLE 5.2. MEASUREMENT AND OTHER UNIFYING IDEAS

| Grade Band | Major Unifying Idea | Problem | Connections with Other Unifying Ideas |
|---|---|---|---|
| 6–8 | Measurement | Squareness problem: A new housing subdivision offers rectangular lots of three different sizes:<br>1. 75 feet by 114 feet<br>2. 455 feet by 508 feet<br>3. 185 feet by 245 feet<br>If you were to view these lots from above, which would appear most square? Which would appear least square? Explain your answers. In what attribute or characteristic of the lots are we interested? In what ways can this amount be quantified? (Simon and Blume 1994) | Quantity<br>Problem solving<br>Geometry<br>Representation<br>Reasoning and proof<br>Data analysis |

As we mentioned in the instructional example for prekindergarten through grade 2, the task itself does not give students the opportunities to make connections among unifying ideas; rather, the thoughtful interactions among the teacher, task, and students does so.

## CONCLUSION

WE HAVE presented information about two unifying ideas in mathematics and described examples of how teaching those unifying ideas might play out in practice. We have also described the connections that the unifying ideas in these particular examples have to other unifying ideas. Our hope is that teachers continue to ask the questions “What mathematical unifying idea underlies this task?” and “What other unifying ideas are connected to the unifying idea in this task?” By carefully considering the answers to these questions, teachers can help their students understand both the unifying ideas in mathematics and how the unifying ideas are interconnected. Table 5.3 gives examples of additional tasks designed to promote the learning of a unifying idea at particular grade levels, along with other unifying ideas that are connected to it. We believe that when students come to understand that the nature of mathematics resides in the connectedness of its unifying ideas, they will then come to understand and seek out those connections.

TABLE 5.3. OTHER REPRESENTATIVE TASKS DESIGNED TO PROMOTE THE INTEGRATION OF UNIFYING MATHEMATICAL IDEAS

| Grade Band | Major Unifying Idea | Problem | Connections to Other Unifying Ideas |
|---|---|---|---|
| Pre-K–2 | Number and operations | Maria has 147 shells in her collection. She wants to have 200 shells altogether. How many shells does she need to find so that she has 200 in her collection? | Place value<br>Communication<br>Problem solving<br>Quantity<br>Representation |
| | Quantity | See table 5.1. | |
| 3–5 | Algebra | Determine the truth of true-and-false number sentences:<br>$7 + 8 = 15 + 4$; $67 + 33 = 100 + 79 = 179$<br>Solve open-number sentences:<br>$x + 57 = 84 + 56$; $x + y = 7$;<br>$436 - 987 + 987 = x$ (Carpenter and Levi 1999) | Number and operations<br>Place value<br>Communication<br>Problem solving<br>Reasoning and proof |

(*Continued on next page*)

Table 5.3—*Continued*

| Grade Band | Major Unifying Idea | Problem | Connections to Other Unifying Ideas |
|---|---|---|---|
| 3–5 | Geometry | Using the commercial product Polydrons, show children two of the regular polyhedra, such as the cube and the tetrahedron, and ask them to find characteristics of all the regular polyhedra and to find the other three regular polyhedra (Lehrer and Curtis 2000). (Each regular polyhedron is made up of a single type of regular polygon, with the same number of polygons meeting at each vertex. The tetrahedron [4 faces], octahedron [8 faces], and icosahedron [20 faces] are made up of triangles; the cube [6 faces] is made up of squares; and the dodecahedron [12 faces] is made up of pentagons.) | Representation<br>Reasoning and proof<br>Problem solving |
| 6–8 | Measurement | See table 5.2 | |
| 6–8 | Algebra | Jeff and Iliana went on vacation. Jeff took \$350 as spending money and spent an average of \$32 a day. Iliana took \$300. She did not spend any money the first two days, but then she spent an average of \$21 a day. Let $x$ be the number of days since Jeff and Iliana left on vacation.<br><br>Write an expression that will show how much money Iliana has left $x$ days after leaving on vacation. Create a graph that shows the amount of money that Jeff has left compared with the number of days he has been on vacation. On the same graph, show the amount of money Iliana has left compared with the number of days she has been on vacation. What does the equation $350 - 32x = 254$ mean in terms of this situation? | Representation<br>Communication<br>Quantity |

(*Continued on next page*)

Table 5.3—*Continued*

| Grade Band | Major Unifying Idea | Problem | Connections with Other Unifying Ideas |
|---|---|---|---|
| 9-12 | Proba-bility | The ELISA test was introduced in the mid-1980s to screen donated blood for the presence of AIDS antibodies. When antibodies are present, ELISA has a positive result about 98 percent of the time. When the blood is not contaminated with antibodies, the test gives a positive result about 7 percent of the time. About .1 percent of the units of donated blood have AIDS antibodies. What percent of the positive results will be false positives? (Moore 1990) | Quantity<br>Problem solving<br>Representation<br>Reasoning and proof |
| | Algebra | Take a sheet of paper. Fold it once. How many layers do you have after 1 fold? After 2 folds? After3? After 4? After 5 folds? If I give you a number of folds, can you tell me how many layers you will have? Tomorrow I will give your friend a problem to determine how many layers result after a particular number of folds. You are to give your friend directions so that he or she can solve the problem. What directions would you give? Describe how many layers you would have after $n$ folds. | Representation<br>Quantity<br>Measurement<br>Problem solving<br>Communication |

## REFERENCES

Carpenter, Thomas P., Elizabeth Fennema, Penelope L. Peterson, Chi-Pang Chiang, and Megan Loef. "Using Knowledge of Children's Mathematics Thinking in Classroom Teaching: An Experimental Study." *American Educational Research Journal* 26 (1989): 499–531.

Carpenter, Thomas P., and Linda Levi. "Developing Conceptions of Algebraic Reasoning in the Primary Grades." Paper presented at the annual meeting of the American Educational Research Association, Montreal, April 1999.

Fennema, Elizabeth; Thomas P. Carpenter, and Megan Loef Franke. "A Longitudinal Study of Learning to Use Children's Thinking in Mathematics Instruction." *Journal for Research in Mathematics Education* 27 (July 1996): 403–34.

Fey, James T. "Quantity." In *On the Shoulders of Giants: New Approaches to Numeracy,* edited by Lynn Arthur Steen, pp. 61–94. Washington, D.C.: National Academy Press, 1990.

Hiebert, James, and Thomas P. Carpenter. "Learning and Teaching with Understanding." In *Handbook of Research on Mathematics Teaching and Learning,* edited by Douglas A. Grouws, pp. 65–97. New York: Macmillan Publishing Co., 1992.

Lehrer, Richard, and Carmen Curtis. "Why Are Some Solids Perfect? Conjectures and Experiments by Third Graders." *Teaching Children Mathematics* 6 (January 2000): 324–29.

Moore, David S. "Uncertainty." In *On the Shoulders of Giants: New Approaches to Numeracy,* edited by Lynn Arthur Steen, pp. 95–137. Washington, D.C.: National Academy Press, 1990.

Simon, M. A., and Glendon W. Blume. "Building and Understanding Multiplicative Relationships: A Study of Prospective Elementary Teachers." *Journal for Research in Mathematics Education* 25 (November 1994): 472–94.

Sowder, Larry. "Children's Solutions of Story Problems." *Journal of Mathematical Behavior* 7 (December 1988): 227–38.

*Teaching to the Big Ideas. Developing Mathematical Ideas, Module 2: Making Meaning for Operations.* [21 minutes ] Newton, Mass.: Education Development Center, 1997. Videocassette.

Thompson, Patrick W. "The Development of the Concept of Speed and Its Relationship to Concepts of Rate." In *The Development of Multiplicative Reasoning in the Learning of Mathematics,* edited by Guershon Harel and Jere Confrey, pp. 179–234. Albany, N.Y.: State University of New York Press, 1994.

# 6

# The Growth of Ability in Data Collection, Organization, Presentation, and Interpretation from Kindergarten through Grade 12

Albert P. Shulte

BEFORE WE consider the big idea of data analysis—the major theme of this chapter—we should examine the role of big ideas in mathematics education and why they should be described in a publication on integrated mathematics. A relatively small number of big ideas are studied throughout schooling. Students return to these big ideas repeatedly, with their understanding deepening over time and through many experiences with the ideas. In addition, these ideas interface with one another as students learn more mathematics and more about the mathematics that they have learned. Each major document on mathematics education takes a slightly different approach to what constitutes the fundamental big ideas. However, most people generally agree that the following ideas should be included:

- Number sense—the ability to make appropriate selections of numbers and operations and to tell whether an answer is correct, for a calculation, or in the right range, for an estimate
- Computational skill—the ability to use mathematical operations and other procedures to reach answers efficiently and accurately
- Measurement—the ability to select appropriate units of measure for a task and to use measurement tools

- Geometry—the ability to work with shapes and shape relationships
- Algebra—the ability to use variables and work with algebraic expressions and algebraic sentences
- Data analysis—the ability to deal with, and make sense of, data
- Probability—the ability to deal with chance events and variation
- Patterns and functions—the ability to see and use patterns and to use particular patterns relating variables

As stated earlier, these big ideas are interconnected. Geometry can be used to visualize algebraic relationships (e.g., using coordinate graphing or algebra tiles to represent binomials), analyze data distributions, visualize patterns, and represent solutions to problems pictorially. Measurement can be used to quantify geometric relationships. Patterns can be represented algebraically, and relationships among numbers can be generalized by using algebra. One way to think about integration in mathematics is to investigate these big ideas and to explore how they are connected.

## DEALING WITH DATA—A BIG IDEA

ONE BIG idea in mathematics education is learning how to deal with data to answer important questions. In the past twenty years, much attention has been given to this topic. To deal effectively with data, one needs to learn, develop, and internalize—

- sound methods for collecting data;
- efficient ways to organize the data collected;
- a variety of ways to present the data once organized;
- appropriate techniques for interpreting the data; and
- meaningful inferences based on the data.

With these skills, students can answer the important questions mentioned above. Furthermore, data collection, organization, presentation, and interpretation should always be done with an eye to seeking answers to relevant questions.

I emphasize the questions because we can easily fall into the trap of concentrating solely on the techniques of using data—how to tally, how to sample a population, how to make a table, how to draw a variety of graphs, how to read and extract information from the tables and graphs, and how to make interpretations—without remembering that the reason we have done all these things is to answer our questions.

This emphasis on remembering the underlying questions has a parallel in our computational studies in mathematics classes. Ultimately, we want to use our computational abilities to solve problems (often non-routine, at least to the person encountering the problem) or to make routine applications of well-known procedures. To achieve these ends, we need to learn basic computational facts, know appropriate and efficient computational algorithms, select appropriate computational tools, and learn problem-solving strategies. However, the ability to solve problems and to apply computation is more than the sum of these separately learned parts. We must implement these skills and techniques in an interactive, thoughtful way—not operate out of rote or habit.

The remainder of this chapter looks at how the ability to deal with data is learned and how it grows as students move from early elementary school through high school.

## WORKING WITH DATA IN EARLY ELEMENTARY SCHOOL

IN EARLY elementary school, children explore basic ideas in mathematics. Thus, children must see their mathematics activities as allowing them to answer questions and relate mathematics to their world.

Most new elementary textbook series do an excellent job of exposing children to a variety of activities involving data collection, organization, and presentation. Children are led to keep track of data by tallying, often in the format of a table. They explore such questions as the following:

- What kinds of pets do the children in this class have?
- How many teeth have most children in the class lost?
- Which of these objects float in water, and which do not?
- What are the favorite television shows of the children in this class?
- What are the favorite foods of the children in the class?

Data are collected from the class—and sometimes compared with data from other classes—and students show amounts by tallying. The first stage in organizing data involves recording the tally information in lists or tables. Once students are comfortable recording data in this way, they can learn to convert the tally sums to numerals.

Once the data have been organized, children can display the information using a variety of graphs. They should move through the sequence of concrete (object) graphs, pictorial (picture) graphs, and abstract (symbolic) graphs (Choate and Okey 1981). Concrete graphs use actual objects on graph grids, for instance, left shoes to answer the question "How are shoes in this class fastened?"; pictorial graphs use pictures of the objects, for example, drawings of the sun, clouds, rain, or snowflakes when weather is graphed over time; and abstract graphs that are appropriate for early elementary students use dot or line plots (NCTM 2000, p. 111).

At this point, young students can begin to explore inferential uses of statistics. Many of us encountered inferential statistics for the first time as part of a formal statistics course in college or graduate school. We might hesitate to think of teaching inference skills to young children. However, the seeds for formal inference need to be sown with young students. Those young students can begin to use the graphs that they have prepared to answer their initial questions. They can also begin to consider whether their class's data might be good predictors for their grade, school, and community or for the country as a whole. This consideration

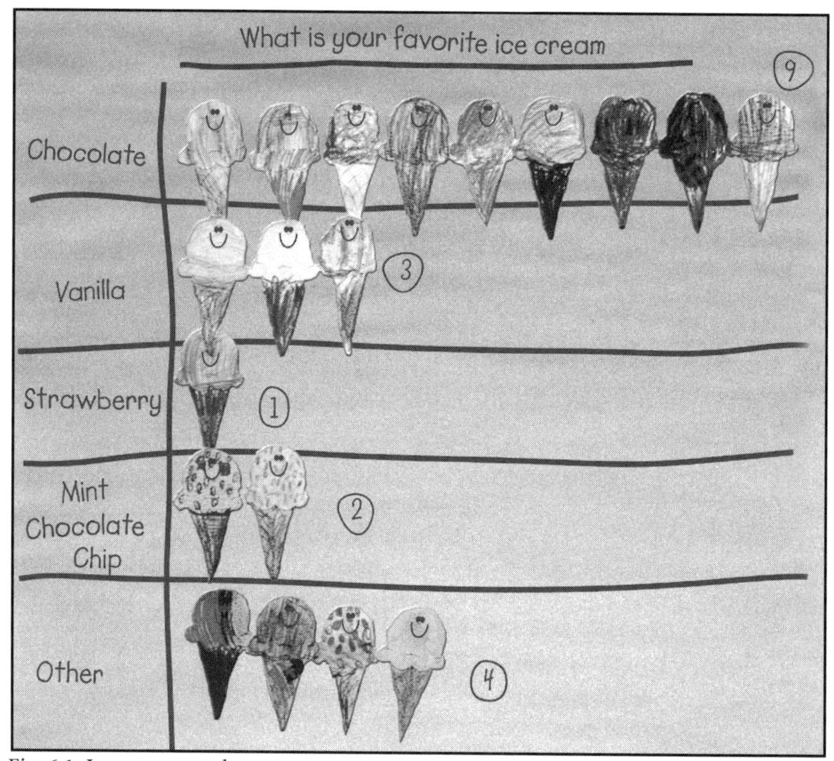


Fig. 6.1. Ice-cream graph

can lead to informal discussions about sampling (e.g., "Of course, everyone in the graph plays a musical instrument. We got the data from the school orchestra").

See figure 6.1 for an example of a pictorial graph of ice-cream preferences prepared in Margaret Risser's kindergarten class at Riverside Elementary School, Waterford, Michigan. The figure shows the flavors considered and has an "other" column for additional flavors preferred. Some good questions would be the following:

- Which flavor is most popular?
- Did most students prefer the most popular flavor?
- How many more children preferred chocolate to vanilla?

## WORKING WITH DATA IN UPPER ELEMENTARY SCHOOL

IN UPPER elementary school, the level of questions that students raise will be substantially more complex. At this time, students can become more sophisticated in collecting data, and a discussion of methods of sampling is appropriate. For instance, if students want to find out what proportion of drivers drive minivans, they should not collect all their data at the school as students are dropped off or picked up. Also, they should not restrict their data collection to particular times of day. They should design ways to collect data at shopping malls, on major streets, in residential neighborhoods, at sports events and concerts, at the local airport, downtown, in city parks, and so on, and to collect the data at different times of day. Students should also understand why collecting data in so many different places and at different times of day is important.

Surveys are another good source for collecting data. Upper elementary school students can survey members of their school community on a variety of topics, such as the following:

- Do students prefer chocolate milk or white milk at lunch?
- How do students feel about the lunch menus offered at the school?
- What are the favorite sports of the students as compared with the favorite sports of their parents?
- How many people are in each household within their school's attendance area? (Note: this question raises again the need for sampling beyond the school walls, because many families have no children of

elementary school age and thus are not represented by the children at the school; other families could be counted multiple times because several of the family's children attend the school.)

Once the data are collected and organized, students can learn and choose from a variety of graphic forms. At this level, reasonable choices include dot or line plots, bar graphs, line graphs, stem-and-leaf plots, and box plots. Students should graph the data in several ways to see whether important information stands out more prominently in one representation than in another (NCTM 2000, p. 176). Students also should explore the shapes of data distributions and look for data clusters, gaps, and extreme values or outliers. They should learn about summary statistics—the mode and the median are accessible and useful measures for students of this age, and the mean can be calculated easily with a calculator. The summary statistics should include measures of spread, such as the range. Quartiles can be used to indicate spread and the range in which the middle 50 percent of the data fall. Students can learn to draw box plots in two ways—to show the entire distribution or to indicate outliers.

An example of a student-created dot plot (line plot) is shown in figure 6.2, which displays the number of raisins in each of twenty-four 1.5 ounce boxes of a popular brand. Some good questons related to the graph are the following:

- If you got a new box of this brand of raisins, about how many raisins would you expect to find in the box?
- Why do you think the graph has high points in two places?

Upper elementary school students can make inferences at a higher level. As they make these inferences, they should consider whether their data-collection methods are representative of the population from which the data were gathered and what generalizations can reasonably be made on the basis of the data.

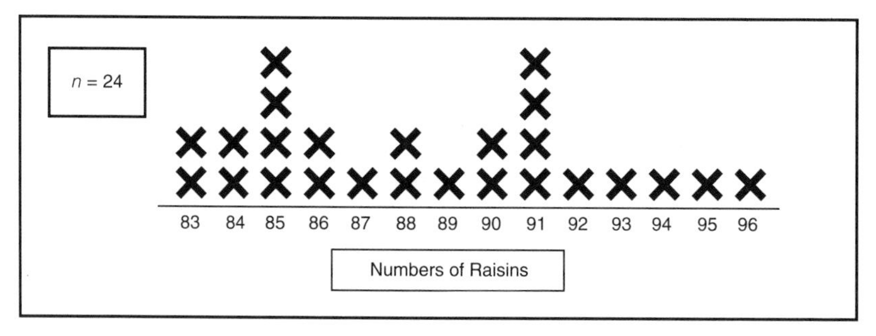


Fig. 6.2. Dot plot showing the number of raisins in each of twenty-four 1.5-ounce boxes of a popular brand.

Classes with easy access to the Internet could collect survey data from other schools (for example, a California school might trade data with one from Maine) or other countries (for instance, at the Fifth International Conference on the Teaching of Statistics, a presentation was made by teachers from Australia and England whose students conducted essentially the same survey—with some wording changes—and discussed their conclusions).

## WORKING WITH DATA IN MIDDLE SCHOOL

IN MIDDLE school, investigating several types of sampling can broaden data-collection techniques. The importance of *simple random* sampling should be stressed, because the procedure offers an equal chance for each individual in the population sampled to be selected. Next, students can consider *stratified* sampling, where the sample should include important categories in approximately the proportion that they appear in the population sampled. It might be important to include different ethnic groups or political persuasions in roughly the proportion that they represent in the population, but selection should then be randomized within an ethnic group or political persuasion. Students can also explore *cluster* sampling, where groups of the population are selected randomly. Census data collected on a sample basis are often gathered using cluster sampling—communities are selected at random, census tracts are selected at random within the communities, and then city blocks are selected at random within the census tracts. In a middle school laid out in a "pod" organization, cluster sampling could involve selecting one to two classrooms from each pod.

To analyze their data, middle school students can employ any of the graphic forms previously learned. They can also be introduced to the scatterplot, where two items of data are collected from each individual in the sample. This practice leads to an exploration of the degree of relationship between the two variables, and this relationship can be studied using lines of best fit to study trends.

Linear trend lines for scatterplots can be developed in three ways. First, one can use the *black thread* technique that has been described to the author by statistician Fred Mosteller, that is, eyeball the scatterplot, see the overall trend, and place on the graph a black thread that approximately fits the trend. Different students can choose different locations and discuss which one seems to reflect the data most closely.

The second way is to use the median-median-fit technique, which is now included on many commercially available graphing calculators.

However, the technique is better learned using paper and pencil first. The steps are as follows:

1. Divide the scatterplot as equally as possible into three sets.
2. In each third of the plot, find the point that represents the (median $x$, median $y$) position.
3. Connect these points in the leftmost third and the rightmost third. This step gives the slope for the line.
4. Look at the (median $x$, median $y$) point in the middle third. Is it close to the line? If so, a linear fit is reasonable. If not, reject the notion of a linear fit.
5. If a linear fit is reasonable, slide the line—maintaining the slope—one-third of the distance from the line to the (median $x$, median $y$) point. This step gives each of the three points equal weight in determining the location of the line.

The third way for developing linear-trend lines for scatterplots involves using a graphing calculator to give a *best-fit* line. In middle school, it is sufficient for teachers to simply indicate that this best-fit line satisfies mathematical principles. Once a student takes first-year algebra, the best-fit line can be revisited and discussed as a least-squares, best-fit line. Students with sufficient algebraic experience can begin to explore lines of best fit in data distributions that are roughly quadratic or exponential and can generate the lines of best fit, or *regression lines*, using graphing calculators. Figure 6.3a presents a table found in a publication from the Pittsburgh International Airport. It lists the flight time, distance, and drive time from Pittsburgh to various U.S. cities. The relationship between flight time and drive time can be investigated by making a scatterplot of the data (see fig. 6.3b) and fitting a least-squares regression line (see fig. 6.3c). The fit is excellent. Some good questions related to the line of best fit are the following:

- If the drive time was 25 hours, about how long would the flight time be?
- If the flight time was 4 hours, about how long would the drive time be?
- About how much flight time is represented by each hour of drive time?

DISTANCE CHART FROM PITTSBURGH INTERNATIONAL AIRPORT

| City | Approximate Flight Time | Approximate Distance | Approximate Drive Time |
|---|---|---|---|
| Baltimore | 1 hour | 218 miles | 4.5 hours |
| Boston | 1.5 hours | 593 miles | 10 hours |
| Chicago | 80 minutes | 452 miles | 8 hours |
| Cincinnati | 1 hour | 295 miles | 6 hours |
| Cleveland | 40 minutes | 129 miles | 2.5 hours |
| Dallas | 3 hours | 1228 miles | 23 hours |
| Erie | 40 minutes | 126 miles | 2.5 hours |
| Los Angeles | 5 hours | 2445 miles | 48 hours |
| Miami | 3 hours | 1168 miles | 21 hours |
| New York | 1.25 hours | 368 miles | 7 hours |
| Omaha | 2.5 hours | 891 miles | 16 hours |
| Philadelphia | 1 hour | 295 miles | 6 hours |
| Phoenix | 4.5 hours | 2118 miles | 40 hours |
| San Francisco | 5.5 hours | 2587 miles | 50 hours |
| Seattle | 5 hours | 2006 miles | 44 hours |
| Saint Paul | 2.25 hours | 849 miles | 15 hours |
| Toronto | 1.25 hours | 324 miles | 6 hours |
| Washington, D.C. | 1 hour | 221 miles | 5 hours |

Fig. 6.3a. Distances from Pittsburgh International Airport to selected cities

Another graphic form appropriate for middle school is the histogram. This form is a natural extension of the bar graph, with the additions that the area under the graph is important and that the data bars are graphed to illustrate ranges of values with consecutive adjacent bars.

An additional possible graph at the middle school level is the circle graph or pie chart. This graph is extremely popular in the news media for expressing parts of a whole, particularly components of budgets. It gives the illusion of being easy to understand and interpret. However, visually comparing sectors that are close in size is difficult. For this reason, most statisticians would prefer that it not be taught. Yet existing practice is sufficiently well established that circle graphs will likely be used in practical situations for many years to come, and thus some exposure to such graphs is desirable.

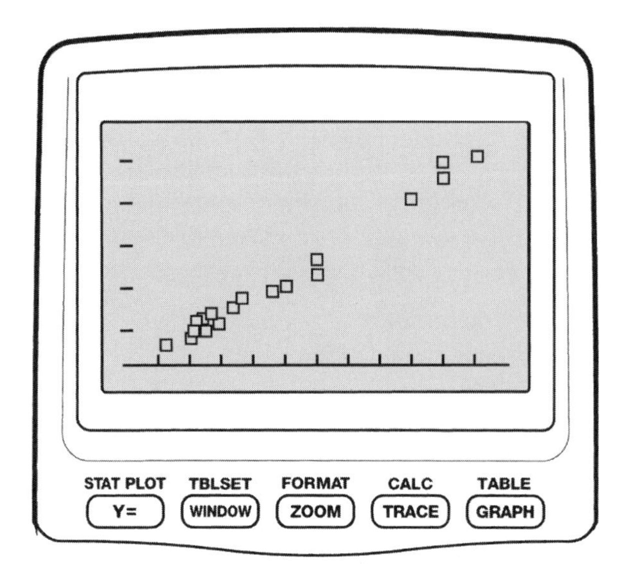


Fig. 6.3b. Scatterplot of the flight-time-versus-drive-time data from Pittsburgh International Airport to selected cities

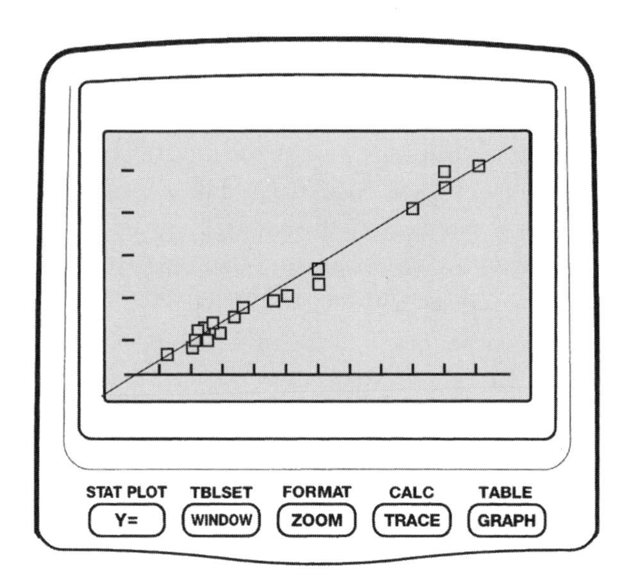


Fig. 6.3c. Scatterplot fitted with least-squares linear regression line of the flight-time-versus-drive-time data from Pittsburgh International Airport to selected cities

Occasionally, some characteristic does stand out when circle graphs are used. One striking example is the relative proportions of blood types (A, B, AB, O) in the populations of various ethnic groups (i.e., Caucasian, African American, Native American, Asian American ). This phenomenon is illustrated in figure 6.4, which is based on a graph that originally appeared in *USA Today*.

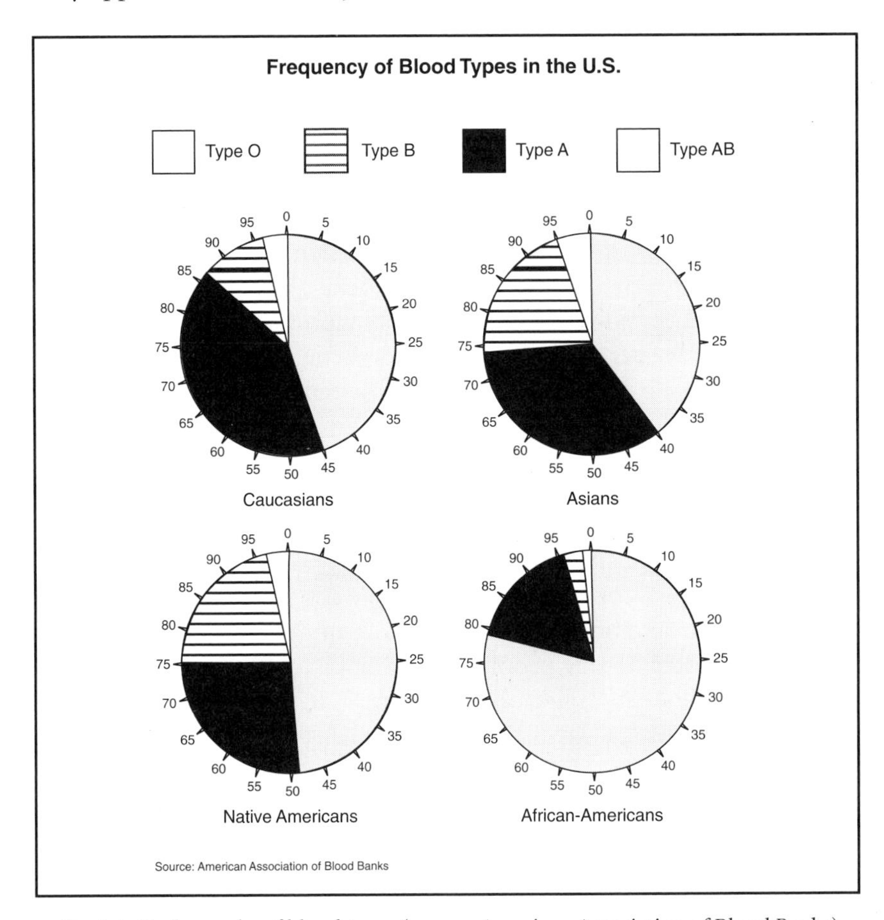


Fig. 6.4. Circle graphs of blood types (source: American Association of Blood Banks)

At the middle school level, data distributions can be much more quantified. A distribution can be characterized using measures of central tendency and spread. Some natural pairings are the median and the range, the median and the quartiles (providing the middle 50 percent of a distribution), the mean and the range, and the mean and the standard deviation. If the standard deviation is used, students should be shown the geometric interpretation of the standard deviation with respect to the normal curve (the *flex point*). They should also learn that roughly two-thirds of the distribution under a normal curve lies within one standard deviation of the mean, that about 95 percent of the normal distribution lies within two standard deviations of the mean, and that approximately 99 percent of the normal distribution lies within three standard deviations of the mean. Students should also learn to inspect graphs of data distributions to decide whether they appear to be approximately normal.

Summary statistics can be calculated using graphing calculators to take much of the tedium out of the calculation. However, students should learn that the mean is the balance point of a distribution, that the median is the middle point of a distribution, and that the mode or modes can be located anywhere within a distribution.

Middle school is not the place for formal inference techniques. However, it is where students should learn to develop generalizations from their data distributions. They should gain experience with making generalizations from one sample (perhaps one group from which survey data have been collected) and with determining whether the generalizations hold for another sample.

## WORKING WITH DATA IN HIGH SCHOOL

DURING HIGH school, students can develop their algebraic experience and skill to use formulas for regression lines. They also should explore the normal distribution and become familiar with confidence intervals. Students can consider the potential problems inherent in collecting data through surveys, observational studies, or experiments (NCTM 2000, p. 326).

By high school, students should be familiar with the entire spectrum of data collection, organization, and presentation devices studied in earlier grades. This preparation enables them to concentrate on data interpretation and making inferences. Students can make assumptions about data distributions and use simulations to see how typical or deviant the samples are.

High school students should build on the modeling that they have done with linear, quadratic, and exponential functions in middle school and use this modeling in a variety of situations. They should also learn to model situations using other functions, such as power functions and logistic functions.

High school students can use techniques of formal inference to investigate questions. In particular, they can make use of the chi-square test and the *t*-test to compare groups, and they can learn which of these tests is more appropriate in a particular situation. For example, they could use the chi-square test to explore whether being left-handed or right-handed affects students' grades in geometry. They could use the $t$-test to compare the effectiveness of two germicidal soaps in disinfecting one's hands. See figure 6.5 for an activity using the chi-square test.

Integrated circuits from two vendors are supplied to a computer manufacturer. Each circuit is tested by the manufacturer for four common defects before it is accepted. The following data, representing test results over a two-week period, have been supplied to the purchasing department by the quality control manager.

| Vendor | Defective Circuits by Type of Defect | | | |
|---|---|---|---|---|
| | 1 | 2 | 3 | 4 |
| A | 60 | 80 | 40 | 30 |
| B | 30 | 32 | 25 | 20 |

At the .01 level, would you accept the hypothesis that the percents of common defects are the same for both vendors? (Source: Sanders, Murph, and Eng [1980])

Fig. 6.5. Exploring the percents of common defects

The defective-circuits problem can be investigated by using the chi-square test for contingency tables. Because it is a relatively small table, it can be calculated by hand. However, the chi-square value and the associated .01 level can easily be calculated using a graphing calculator or a computer statistical package. The null hypothesis is that no difference exists between the percents of common defects for the two vendors; the alternative hypothesis is that a statistically significant difference exists.

On the TI-83 graphing calculator, by using matrix A for the observed data in the table, matrix B, the expected data, is calculated as is the chi-square value of 2.87. With 3 degrees of freedom, the associated probability is 0.41—a long way from the .01 level (which would require a *chi-square* value of 11.345). Hence, one need not assume that the two vendors differ in their percent of defects.

## INTERACTIONS WITH OTHER BRANCHES OF MATHEMATICS, KINDERGARTEN THROUGH GRADE 12

AS STUDENTS learn, grow, and mature in their abilities to collect, organize, present, and interpret data, and as they become increasingly sophisticated in making inferences on the basis of data, they interact with much of the rest of the mathematics curriculum.

In early elementary school, as students work with data, they reinforce their skills in counting and measuring. As they progress through the elementary grades, they use numerical operations, such as addition and division in finding averages, and increase their skills in ordering numbers, for example, when determining medians and quartiles. They also begin to examine the shapes of data distributions, thus making geometric connections. In middle school, scatterplots have obvious connections with the coordinate plane, and lines of best fit require the ability to graph algebraic equations and to use formulas for linear equations. Later in middle school, lines of best fit in nonlinear situations expose students to quadratic and exponential functions. In studying the normal distribution, high school students deal with algebraic and geometric interpretations of standard deviation. When inference tests are used, students increase their abilities to work with formulas. Inference itself extends students' work with logic and problem solving. Throughout the development of ability in working with data, students use growing notions of chance and probability. The discussion of appropriate growth in probability skills, from informal ideas of chance to working with various probability distributions, would require another chapter of comparable length.

The growth in ability to work with data obviously connects students with the world beyond the classroom. The most appealing examples of this connection are drawn from real questions and are dependent on real data. The use of data gathered outside the classroom and the interaction of the analysis of the data with other mathematical studies infuse realism into the mathematics curriculum.

A look at the ten Standards in the appendix of *Principles and Standards for School Mathematics* (NCTM 2000) shows the degree to which data analysis is interconnected with other strands. The Content Standards merge with Data Analysis in the following areas:

- Number and Operation—includes ways of representing numbers (which can involve graphs and formulas), understanding relationships among numbers (clearly a major part of analyzing data), computing fluently (e.g., calculating means and standard deviations), and making reasonable estimates (related to formulating hypotheses and making predictions)
- Algebra—includes the goals of understanding patterns, relations, and functions (related to making sense of data); representing and analyzing mathematical situations and structures using algebraic symbols (e.g., using formulas for the mean, standard deviation, and distributions, such as the normal distribution and the binomial distribution); using mathematical models to represent and understand quantitative relationships (modeling plays a key role in data analysis, display, and statistical inference); and analyzing change in various contexts (certainly the business of data analysis as well as of many other mathematical disciplines)
- Geometry—emphasizes the use of coordinate geometry (used explicitly in graphing data as scatterplots, and used in general in all graphing through the use of coordinate axes), encourages the use of symmetry to analyze mathematical situations (symmetry is one of the items examined when studying graphs), and recommends the use of visualization (a major part of data analysis) to aid in problem solving
- Measurement—fits hand-in-glove with data analysis, since determining the range, quartiles, median, mean, and standard deviation of a distribution are all forms of measurement. Furthermore, graphs (e.g., box plots) are based on these measurements.

Turning from the Content Standards to the Process Standards, data analysis interacts with these strands in the following ways:

- Problem Solving—includes applying and adapting a variety of appropriate strategies to solve problems and reflecting on the process of mathematical problem solving
- Reasoning and Proof—includes making and investigating mathematical conjectures and selecting and using various types of reasoning
- Communication—covers communicating mathematical thinking coherently and clearly to others and using the language of mathematics to express ideas precisely

- Connections—entails recognizing and applying mathematics in contexts outside mathematics
- Representation—involves creating and using representations to organize, record, and communicate mathematical ideas; selecting, applying, and translating among mathematical representations to solve problems; and using representations to model and interpret physical, social, and mathematical phenomena

Clearly, the various mathematical strands are not independent entities; rather, they are major structural parts of mathematical work, and they interact in many ways. Data analysis is enriched by the understanding and use of other mathematical strands and is also an important area of application for the other strands. After all, to truly answer important questions in the real world, we need to draw on all the big ideas in mathematics.

## REFERENCES

Choate, Laura Duncan, and JoAnn King Okey. "Graphically Speaking: Primary-Level Graphing Experiences." In *Teaching Statistics and Probability*, 1981 Yearbook of the National Council of Teachers of Mathematics (NCTM), edited by Albert P. Shulte and James R. Smart, pp. 34–37. Reston, Va.: NCTM, 1981.

National Council of Teachers of Mathematics. *Principles and Standards for School Mathematics.* Reston, Va.: NCTM, 2000.

Sanders, Donald H., A. Franklin Murph, and Robert J. Eng. *Statistics: A Fresh Approach.* New York: McGraw-Hill, 1980.

# Part 2

## Planning, Implementing, and Assessing Curricula for Integrated Mathematics

7

# Team Planning to Create an Integrated Curriculum

John Burrill

Victor M. Hernández-Gantes

IN THE 1990s, one of the problems with U.S. education became clear—high school graduates did not have the occupational skills they needed to function in today's technical workplace. This realization triggered a productive discussion on how mathematics is fundamental in helping individuals succeed in the world of work (Mathematical Sciences Education Board 1995). Concurrently, the School-to-Work Opportunities Act of 1994 supported program initiatives that prepared students to enter the workforce or pursue further education after high school (U.S. Congress, Office of Technology and Assessment 1995). Coming from two different perspectives, both reform movements called for integration and application of academic concepts in occupational contexts. In response to this call, a number of applied curriculum materials have become commercially available, but questions persist regarding their quality and actual implementation. Our experience shows that interdisciplinary teacher collaboration may be the missing link needed to effectively integrate mathematics and technical skills in ways that are relevant for local school communities.

The MathNet project, conducted from October 1995 to September 1997 by staff from the University of Wisconsin—Madison and funded by the U.S. Department of Education, Office of Vocational and Adult Education, offered the opportunity to explore the importance of collaboration in integrated curricula. The project consisted of eight teams of academic and vocational instructors representing seven high schools and a community college. The teams were charged with integrating mathematics and career-oriented curricula, and they were assisted by the University of Wisconsin's Center on Education and Work and the National Center for Research in Mathematical Sciences. This chapter is based on the curriculum integration experiences generated through the MathNet project (Hernández-Gantes, Brendefur, and Burrill 1998).

This chapter outlines two important steps—selecting and building a team and team planning—for facilitating interdisciplinary curriculum development between vocational and mathematics teachers. The chapter also examines how to develop curriculum units that are based on either worthwhile problem situations or on contexts that offer opportunities to develop and use problem-solving skills related to mathematics and vocational-technical curricula. Furthermore, the chapter highlights the design challenges that arise when developing an interdisciplinary curriculum. Those challenges are designing worthwhile problems, adjusting instruction, facilitating collaboration, and designing appropriate assessment strategies.

## SELECTING AND BUILDING A TEAM

WHILE WE were assisting with creating secondary-level curriculum units that integrate occupational contexts and mathematics in the MathNet project, we realized that such units could not be designed by just one individual. Rather, to effectively integrate the curriculum, mathematics and vocational education instructors would need to collaborate.

Creating the team is the first step. Team members should be selected from among educators committed to integrating disciplines, implementing school reform, creating standards-based curricula, and developing collaborative curricula. Note that one should not necessarily choose the senior members of the department or the chairperson, and one should avoid offering team membership as a reward for a job well done or as extra compensation. The individuals selected must be creative leaders who are able to work with others on a team. Members of each discipline need to be very well prepared in their chosen discipline and its applications. To understand and implement a high-quality integrated unit, team members need to have experience in various teaching formats and knowledge of school reform in both mathematics and vocational-technical education.

Next, once the team members have been selected, the team must begin by examining the school curriculum and teaching practices to identify opportunities for improvement. The team needs to build a general understanding of current vocational-technical and mathematics instruction. Some important questions include the following:

- Is vocational instruction mostly show-and-tell?
- Do the students primarily work on projects following step-by-step instructions?
- Would the students benefit from work-based experiences?
- Is mathematics teaching primarily paper-and-pencil drills?
- Is rote memorization encouraged to help students learn formulas and steps to solve problems?
- To what extent are the students applying mathematical concepts in real-world situations?

The answers to these and other relevant questions will help the team understand current practices. Next, the team should create a frame of reference for change and then agree on the nature of the change. For example, the concepts of *critical thinking*, *teaching for understanding*, and *problem solving* are used in school-reform language. The team members need to agree on what these concepts mean and how they would put them into practice. This agreement is essential for effective team building and for successfully implementing any curriculum, integrated or not.

## TEAM PLANNING

THE MATHNET project confirmed that the second important step for effective curriculum integration was team planning. The team needs significant time to reflect on the scope and nature of its efforts to integrate mathematics in occupational contexts. This reflection gives them a better opportunity to build an interesting project that also fits in the school curriculum.

For effective planning, the team must agree on what they want to accomplish as a school and as a team as they integrate mathematics and vocational-technical content. The team's vision should articulate how the curriculum changes will improve the students' learning (Kruse, Seashore, and Bryk 1994). A great deal of planning time is needed to develop this shared vision. Teacher-teams are often surprised at how much discussion is needed to achieve a consensus on language and reasons for integrating mathematics and occupational contexts.

As part of refining its vision, the team must develop and agree on specific criteria to guide the common vision. These criteria are needed to explain to students and others what is meant by "high-quality teaching," "challenging student learning," and other jargon that may warrant clarification.

Teams should also identify why they are coming together to create an integrated curriculum. One reason might be a school mandate to create applied courses. In other instances, the reason might be to develop an applied project that will connect the students with problems encountered in occupational contexts. Or, the reason fueling team efforts could be to plan an integrated curriculum to support a career academy. At any rate, identifying the main reason for integration allows the teams to gauge the scope and nature of the curriculum to be planned. An applied course, for instance, may contain integrated concepts but may not require team teaching; conversely, integrated curricula for a career academy could require team teaching. Similarly, applied projects may take the students outside the school to document solutions but may not require work across courses.

Collectively, team planning develops rapport and reinforces common goals. In most instances, teachers rarely have an opportunity to collaborate with peers outside their departments. But without planning time to talk and collaborate, the development of a standards-based integrated unit would be very difficult. Vocational-technical teachers may not understand the complexity of mathematics applications, and mathematics teachers may not grasp the practical implications in occupational contexts. Once the team members become aware of their planning and collaboration needs, they should have a better idea of the challenges that they face.

## DESIGNING WORTHWHILE PROBLEMS

ONCE A working interdisciplinary team is established and begins planning, the members face the challenging task of designing a worthwhile problem scenario for the students. The problem should be grounded in an occupational context relevant to students and instructors and should fit into the school curriculum. Integrated curriculum units must meet course requirements for each discipline. The team should also realize the unreasonableness of thinking that all learning should be integrated. Any attempt to base all instruction in occupational contexts would be very difficult—and in many situations meaningless. The focus of the unit must be on the essential concepts from each discipline that can be addressed equitably throughout the unit.

Consider the approach taken by a three-member team of mathematics, physics, and technology instructors from River Dell High School in Oradell, New Jersey. The team determined that a connection existed within their three disciplines and that the connection could be applied through a unit on bridge safety. Here is the scenario that they presented to their students (Ciccotelli, Giglio, and Piekielik 1998, p. 3):

> A pedestrian bridge is needed at your local amusement park. The bridge needs to span a small river approximately 30 meters wide. Local engineers have decided to accept designs from local schools for the 'best' bridge design. The bridge design must guarantee safety and efficiency. A scale model and drawing must be submitted.

In this example, the team identified concepts and tasks from each subject area. From mathematics, the concepts included linearity, direct and inverse variation, parabolic functions, and properties of triangles. From physics, they identified vector representation of a static force, the conceptual model of static equilibrium by identifying balanced opposing forces, and the resolution of a force into its components. The technology class, which was called junior engineering, gave students the opportunity to select, adapt, use, and manipulate materials to fabricate a prototype for each design and to test and evaluate the prototype according to the problem's requirements. The technology class also required understanding the use of tools and techniques needed by the design engineer and the safety procedures needed in the fabrication of prototypes (Ciccotelli, Giglio, and Piekielik 1998).

This example shows that team members have to plan together and refrain from simply pasting one subject onto the other. At first, team members may want to take an existing unit in their discipline, determine mathematical or occupational context items, and implement that preexisting unit rather than collaborate on the design of a new one. At the heart of such an approach is the fact that neither party may have the knowledge necessary to determine whether the mathematics or occupational items they identified are relevant to the other instructor's curriculum. The team needs to review and share existing curricula, search for possible complementary concepts, and then select a subset of those concepts around which they could design a unit. This process is a continuation of the team planning discussed previously. The goal is not to identify "good" or "neat" projects but rather to build a problem requiring the use of important mathematical concepts interwoven into a broad occupational context (Hernández-Gantes, Brendefur, and Burrill 1998).

Collaborative curriculum development is not an easy task. It requires interdisciplinary collaboration among teachers and involves creating standards-based learning featuring rigorous mathematics in occupational contexts. Such interdisciplinary collaboration may be foreign to most teachers and administrators. It challenges well-entrenched beliefs about the value of mathematics teaching and learning, vocational-technical education, and the role of the instructors and students (Hernández-Gantes and Brendefur 1997).

From the MathNet project, we saw the greatest success when problem scenarios were anchored in real-world situations and featured solutions that used selected concepts. Strategies for developing problem scenarios include (a) designing a problem using brainstorming techniques and on the basis of the experiences of team members, (b) surveying local industries, employers, and professionals in the community to identify important occupational integration possibilities, and (c) seeking workplace-learning experiences for teachers with local employers, community agencies, or government agencies. Of those strategies, the surveys and actual workplace experiences appear to lead to meaningful and innovative service-learning projects for students (Sargent and Ettinger 1998).

To further refine the problem scenario, it is necessary to determine whether the occupational context is also worthwhile. Possible questions include the following:

- Will the problem allow the students to use knowledge and relevant vocational and technical skills?
- Will the students be able to apply and understand the value of important mathematics concepts in real-world situations?
- Will the problem ensure the students' active participation in a variety of activities and include hands-on learning?

Consider how an instructional team from Swansea High School in Swansea, South Carolina, addressed these questions. The team wanted to help students establish conceptual connections between mathematics and automotive technology. Because the majority of an automobile's functions rely on electronics concepts and mathematical interpretations, the team decided that the connections were ideal for an integrated unit. They developed the problem scenario in the context of troubleshooting electrical and electronics systems in automobile repairs. This troubleshooting context provided rich opportunities for students in grades 9 through 11 to apply both mathematics and technical concepts as they attempted to solve real automotive problems.

To solve the problem, the students—who were enrolled in both applied mathematics and automotive technology—had to understand and use the ideas of graphs, patterns, and equations to identify, diagnose, and repair electrical systems in automotive technology. Specifically, the students measured resistance using a digital voltmeter, learned the color-coding of resistors, and discovered the characteristics of a diode. Concurrently, the students used graphs and schematic diagrams to discuss AC/DC circuits, explained capacitors, investigated the action of a variable resistor, and used the concept of independent and dependent variables to describe settings on the equipment and graphs. With the aid of graphing calculators, the students graphed and symbolically represented alternating and direct current; experimented and gathered data to determine the relationship among current, voltage, and resistance; and investigated direct and indirect variation and the relationship between an equation and its graph through Ohm's law. In this example, the team based the unit on the premise that to troubleshoot and repair an automobile's electrical and electronic circuits effectively, an integrated understanding of automotive technology and mathematics would be essential (Hickman and Jowers 1998).

The identification of shared objectives and complementary concepts should focus the efforts of any interdisciplinary team and guide related work toward developing an integrated curriculum. The task of addressing a whole occupational context at one time is too difficult, so breaking the context down into themes or subsets of concepts is important to getting started.

## ADJUSTING INSTRUCTION AND FACILITATING COLLABORATION

THE NEXT challenge for an interdisciplinary team is to design teaching strategies and collaborative structures. A shift in teaching and collaboration among teachers must occur. Creating integrated units should be a joint effort, and the teaching, as suggested by the MathNet project experience, needs to become team teaching. Teams should work with school administrators to facilitate the assignment of two teachers to a class of students involved in the project of interest. Or, alternatively, the team could negotiate the assignment of a preparation hour or nonteaching assignment (i.e., study hall monitor) to one team instructor at the same time as the integrated class is to be taught. This second method enables the teachers to trade assignments from time to time so that the teacher best prepared for a given section can be in the classroom at the appropriate time.

Team members should discuss and agree on ways to design effective teaching strategies that depart from traditional lectures and paper-and-pencil exercises. Integrated instruction calls for active learning, data collection, and demonstration of potential solutions in school and real-world settings. Because of these requirements, the traditional lecture- or textbook-driven lessons should not be the main teaching method. The team should also be aware that integrated instruction requires additional resources, such as materials, equipment, and access to facilities and human resources from both the school and community. Some of these additional resources cost money, and the instructional design must be based on the funds available. Before team members move forward with their planning, they must be sure that all the materials and equipment needed to teach an integrated unit are available.

## DESIGNING APPROPRIATE ASSESSMENT STRATEGIES

ONCE THE team is satisfied with the final version of the problem scenario, the next challenge is to design appropriate assessment strategies. The team should answer the following questions:

- Within the context of the scenario, what should the students be able to do?
- How will the students demonstrate their results?
- What will be the criteria for evaluating students' work?

One approach that a team could follow is to establish a list of all the mathematical and technical knowledge and skills that the students need to solve the final problem. In creating the list, the team should consider that embedding high expectations enhances the learning by all students. Once the list is created, it should be translated into outcomes and a format in which results can be demonstrated. For example, one of the expectations might be that within the production of an item, the total cost must be analyzed. The results could be demonstrated using a spreadsheet, a written report, or some combination thereof. Most important, students must possess multiple ways to demonstrate their knowledge and skills.

The team should also be aware, however, that creatng the list of expected outcomes from the problem is not sufficient for meaningful assessment. The team has to develop criteria for monitoring progress equitably. The criteria must be appropriate and specific for each of the student expectations, and the team must communicate the criteria to the students at the outset. For example, the team may decide that the students will have to formally present their results. An important question for the team to address is "What should be the minimum requirements for this presentation?"

Teacher-teams must keep in mind that the lists of concepts and skills cover only the major outcomes from the problem scenario. The teams also need to create a more comprehensive assessment plan for continuous feedback to the instructors. Teacher-teams can create contracts or rubrics to help both the students and the instructors in the day-to-day assessment. For example, see figure 7.1, which shows the sample rubric created by the MathNet team from Swansea High School for their unit, "Discovering and Applying Mathematics and Automotive Technology" (Hickman, Jowers, and Sarvis 1998). This rubric appropriately establishes the criteria for assessment and describes various levels of acceptance. Not included in the figure, but a part of the assessment, was a set of graded papers that also illustrated the expectations.

| Content | Novice | Technician | Master Technician |
|---|---|---|---|
| • Effective use of Ohm's law<br>• Achievement of acceptable voltage drop values in automotive circuitry<br>• Calculation of resistance from the voltage drop<br>• Understanding of automotive starting system circuitry<br>• Effective use of graphs<br>• Use of supporting arguments | • Shows gaps in knowledge; misunderstands major ideas and concepts; may fail to include relevant content<br>• Fails to justify response or offers weak argument | • Shows knowledge of major ideas and concepts; covers required content<br>• Justifies response with adequate detail | • Shows clear understanding of major ideas and concepts; fully covers required content; explains how ideas and concepts apply to the scenario<br>• Convincingly justifies response with well-developed reasoning and detail |
| **Analysis** | | | |
| • Address scenario requirements<br>• Evaluation of evidence | • Does not address the scenario<br>• Ignores evidence or demonstrates incomplete understanding of the presented problem | • Addresses all scenario requirements<br>• Links response to evidence | • Addresses all scenario requirements in detail<br>• Links response to all evidence presented |

| Communication | Novice | Technician | Master Technician |
|---|---|---|---|
| • Organization and clarity<br>• Language mechanics | • Ideas are presented in a disorganized way<br>• Writing style or mechanics interfere with communication of ideas | • Ideas are presented in an organized way<br>• Ideas are understandable; language errors do not interfere with communication | • Writing is clear and well organized throughout the response<br>• Ideas are presented effectively and are easy to understand |

Fig. 7.1. Written scenario rating guide (from Hickman, Jowers, and Sarvis [1998])

# CONCLUDING REMARKS

INTEGRATING MATHEMATICS and vocational and technical concepts is exciting and challenging. It requires teachers and their home schools to commit to the effort. Successful projects require interdisciplinary teacher-teams engaged in collaborative, integrated, curriculum development. They also require careful selection of participants, ongoing planning, and a feasible idea of how much the team can accomplish given the existing school curriculum and available resources. Finally, schools must commit to team efforts and offer the necessary structural supports, such as planning time, to help the teams create powerful teaching and assessment strategies that support relevant mathematics learning for all students.

## REFERENCES

Ciccotelli, Tony, Louis Giglio, and John Piekielik. "Bridges: How Safe Are They?" In *Integrating Mathematics in Occupational Contexts: A Sampler of Curricular Units,* edited by Victor M. Hernández-Gantes, Jonathan Brendefur, and John C. Burrill, pp. 57–66. Madison, Wis.: Center on Education and Work, University of Wisconsin—Madison, 1998.

Hernández-Gantes, Victor M., Jonathan Brendefur, and John C. Burrill. *Integrating Mathematics in Occupational Contexts: A Sampler of Curricular Units.* Madison, Wis.: Center on Education and Work, University of Wisconsin—Madison, 1998.

Hernández-Gantes, Victor M., and Jonathan Brendefur. "Producing Authentic, Integrated, Standards-Based Mathematics Curriculum: [More Than Just] a

Collaborative Approach." Paper presented at the annual meeting of the American Educational Research Association, Chicago, March 1997.

Hernández-Gantes, Victor M., Jonathan Brendefur, and John C. Burrill. *Integrating Mathematics in Occupational Contexts: A Guide for Developing Integrated Mathematics Curriculum Units.* Madison, Wis.: Center on Education and Work, University of Wisconsin—Madison, 1998.

Hickman, Heyward, Sam Jowers, and Sandra Sarvis. "Discovering and Applying Mathematics and Automotive Technology." In *Integrating Mathematics in Occupational Contexts: A Sampler of Curricular Units,* edited by edited by Victor M. Hernández-Gantes, Jonathan Brendefur, and John C. Burrill, pp. 67–78. Madison, Wis.: Center on Education and Work, University of Wisconsin—Madison, 1998.

Kruse, Sharon, Karen Seashore Louis, and Anthony S. Bryk. "Building Professional Community in Schools." *Issues in Restructuring Schools,* Report 6 (1994): 3–6.

Mathematical Sciences Education Board. *Mathematical Preparation of the Technical Workforce: Report of a Workshop.* Washington, D.C.: National Academy Press, 1995.

Sargent, Thomas, and Judith Ettinger. "Educator Internship Programs: Providing a Quality Learning Experience." In *Teacher Learning in Workplace and Community,* edited by L. Allen Phelps, pp. 23–34. Madison, Wis.: Center on Education and Work and National Center for Research in Vocational Education, 1998.

U.S. Congress, Office of Technology and Assessment. *Learning to Work: Making the Transition from School to Work,* OTA-HER-637. Washington, D.C.: U. S. Government Printing Office, 1995.

# 8

# Studio Calculus and Physics: Interdisciplinary Mathematics with Active Learning

Karen Marrongelle
Kelly Black
Dawn Meredith

NEWTON DEVELOPED calculus so that he could elegantly explain the motion of the moon around the earth and that of the earth around the sun. Since the time of Newton, calculus and physics users have developed deep connections between the two disciplines. Yet today, these connections are obscured even for many students who take both courses. The reason for this obscurity is perhaps calculus and physics courses are offered by separate academic departments and by different teachers who communicate little, if at all. Why do we teach connected ideas in a disconnected way? Perhaps because teachers assume that students can make the connections themselves or because running a combined course seems difficult or not worth the effort.

Making connections between calculus and physics is difficult for students because of the differences in the courses—including notation, goals, vocabulary, and schedules—and the students' own inexperience with both subjects. For example, students may be given an algebraic derivation of $v(t) = v_0 + at$, where $a$ is the constant acceleration and $v$ is the velocity. Two months later when they learn to integrate, are these students likely to realize that integration is simply a special case of $v(t) - v_0 = \int_0^t a(t')dt'$? Are students who are given the constant acceleration equations going to see the unity of physics and calculus in those equations? Evidence suggests that they do not (Sherin 2001).

Making connections and transferring ideas to a new context are difficult processes that many students cannot accomplish on their own (Brown 1989). For students to transfer an idea to a new subject, they must first abstract that general idea from the few particular situations they have seen. The opportunity to make such connections does not always arise in the context of single-subject classes unless the teachers take time out of their already busy schedules to prompt the connection.

Furthermore, students believe that the information they receive in school is disconnected from a context that is relevant to them, according to reports from the National Science Foundation (1994; 1996) and the National Research Council (1996). Finally, even when instructors agree that teaching calculus and physics should be coordinated, they may not do so for a number of practical reasons.

In this chapter, we present a successful curriculum that combines calculus with physics for first-year college students. This curriculum was developed under the auspices of a grant from the National Science Foundation (grant number DUE-9752485). We found that common goals—studying change and superposition and improving problem-solving skills—were essential to the curriculum's success, as was a revised calculus schedule that allowed the students to immediately apply the calculus that they were learning to the physics that they were learning. We also found that pedagogical issues needed to be addressed. For example, we grounded many of our in-class activities in theories of active learning (Greeno and the Middle-School Mathematics through Applications Project Group 1997) as well as in research that showed the benefits of explicit instruction in problem solving (Schoenfeld 1985). In this chapter, we describe the processes of planning, implementing, revising, and evaluating our combined curriculum.

# PROGRAM DESIGN

## Institutional Context

MOST OF the first-year students from the University of New Hampshire's College of Engineering and Physical Sciences enroll in introductory calculus and physics classes. Those classes meet three days a week in a large-lecture format. In addition, the physics students meet once a week in smaller labs. Both calculus and physics students meet once a week in smaller problem-solving sessions. We refer to these introductory calculus and physics classes as the "regular" calculus class and the "regular" physics class throughout this chapter. Typically, between 100 and 150 students attend each of these lecture classes.

Calculus 1 is a corequisite for physics 1, and calculus 2 is a corequisite for physics 2.

## Planning the Course: Goals

The first task in planning a combined calculus and physics program was to carefully articulate our goals. The process goal was for the students to improve their problem-solving skills; the content goals were for the students to be able to understand and work with change and superposition.

### Problem-solving goal

The general "expert" problem-solving process that we wanted the students to learn has been described similarly in many places: "understand the problem, devise a plan, execute the plan, look back" (Pólya 1957). Yet this process is extremely difficult for learners, who typically skip all the steps except for the execution. Because of this resistance from students, researchers call for direct problem-solving instruction in both calculus and physics classrooms (Schoenfeld 1985; Arcavi et al. 1998). Additionally, Schoenfeld (1983) and Lester and Garofalo (1982) discuss the importance of using metaprocesses, including having the problem solvers monitor their own progress and evaluate their own procedures and results as they solve problems.

In deciding how to promote the use of all four problem-solving steps, we were guided by the work of Heller, Keith, and Anderson (1992), who used cooperative groups and "context rich" problems to facilitate and encourage the use of all four steps. Context-rich questions are closer to real-world problems than are typical textbook problems and share the following four characteristics (Heller, Keith, and Anderson 1992):

(1) The problem statement does not always explicitly identify the unknown variable.

(2) More information may be available than is needed to solve the problem.

(3) Information may be missing, but can easily be estimated or is "common knowledge."

(4) Reasonable assumptions may need to be made to solve the problem.

The researchers found that students appreciate the need for all four steps in problem solving only when the questions have the four features listed above. Therefore, we designed our curriculum so that students worked on context-rich problems throughout the semester and also worked on several longer context-rich problems as periodic projects.

This problem-solving goal is consistent with the approach of combining calculus and physics instruction in two aspects. First, the literature shows that experts in any field have a richly connected knowledge set that helps them in problem solving, whereas naive problem solvers have a sparse, disconnected knowledge set (Glaser 1992; Larkin 1979). By explicitly showing connections between calculus and physics and by looking at general rules instead of particular special instances, we hoped that the students would develop a more richly connected knowledge structure than they would have developed in a typical introductory course and thus would enhance their ability to solve problems.

Second, we wanted the course to give a rich physical context to the mathematics. However, the physics also needed a rich, real-life context. Often physics problems are disconnected from real life. For example, a problem asking students to find the forces on a block that is moving on an inclined plane is largely disconnected from students' everyday experiences and interests. Students can see the connection with real life when they explore such reality-based questions as "How much time will elapse before a comet with a given initial speed and location hits the earth?" "How do neon lights work?" "How does a gas gauge work?" "What types of bicycle wheels are best under what conditions?" (For other context-rich problems, see the Web page of the University of Minnesota Physics Research and Development department at www.physics.umn.edu/groups/physed/ or see the problems from the University of Massachusetts Physics Education Group [Leonard and Gerace 1994]. Our projects can be found at http://www.unh.edu/calculusphysics.)

## Change and superposition goals

Both calculus and physics courses explore how quantities change and how they combine. Therefore, understanding the ideas of change and superposition were natural choices for content goals for both courses. These overall content goals helped us decide on what topics to present. Combined topics included the following:

1. Examining position, velocity, and acceleration in kinematics in conjunction with average and instantaneous rates of change in calculus
2. Exploring total change in force either over time or over space (impulse and work) in physics in conjunction with Riemann sums and integrals in calculus

3. Using the differential equations that arise from Newton's second law ($F = m\ a = m\ dv/dt$) in physics to introduce the mathematical ideas of how to solve differential equations (e.g., slope fields, Euler's method, and solutions of linear differential equations with constant coefficients)
4. Analyzing how exponential and trigonometric functions are generated by physical phenomena that change in time (e.g., the temperature of cooling coffee and the position of a pendulum)
5. Calculating the total electric field due to a bar of charge as the superposition of the electric field due to each charge in that bar
6. Introducing the superposition of Fourier components to describe periodic waves
7. Calculating electric flux through a surface as the superposition of electric flux through each infinitely small flat part of the surface

Although we tried to keep the courses as coordinated as possible, instances arose when the calculus and the physics were not well matched. For example, the study of infinite series in calculus did not have as many applications in physics as the study of differentiation and integration did. When we began a topic that was not coordinated, we pointed this fact out to the students so that they were not looking for connections that were not forthcoming.

## Planning the Course: Schedules and Studio Format

Topic order was also a significant influence in successful implementation. The students needed to see the applicability of the calculus *as they were learning it*, and at the same time, they needed to realize that they were familiar with all the mathematics required to solve the current physics problems. At first the problem of topic coordination seemed insurmountable because often students need the ability to differentiate within the first month of physics, yet the topic of differentiation is not introduced in calculus until a month or two later. However, we overcame the challenge by discussing the four basic threads of calculus—function, continuity, derivative, and integral—first for polynomial functions only and then again for logarithmic, exponential, and trigonometric functions as they arose in the physics curriculum.

This reordering of the calculus curriculum allowed us to present the physics content in a more unified way and gave the mathematics content a rich context. For example, by the end of the first month of the combined class, the students could use antiderivatives to calculate velocity and position as a function of time if they were given an acceleration that was a polynomial function of time and initial conditions. In contrast, the regular physics students spent a good deal of time algebraically manipulating the constant-acceleration equations and often failed to understand that these equations were limited in their applicability.

Beyond the topic sequencing, which was a problem unique to the combined course, we also faced more familiar pedagogical issues. The most basic questions regarded format. For instance, how much do we lecture? How much group work is needed? Although the fundamental ideas could be presented in lectures, experience as well as research had shown us that lecturing alone was not the most effective learning method (Hake 1998; NSF 1996; NRC 1996). Therefore, we adopted the idea of the "studio" format that was pioneered at Rensselaer Polytechnic Institute (Wilson 1994). The format features several elements:

1. Less lecturing by professors and more hands-on work by students
2. Cooperative-group work by students
3. Tightly integrated labs and lectures for the physics portions of the content; labs taking place in the same room as lectures and clarifing and extending the concepts just presented in the lecture
4. Curriculum materials that have been shown through research to be effective

## Program Description

The calculus and physics students met five days a week for one hour and fifty minutes a day. The students met in a physics lab with the physics instructor and teaching assistant two days a week and in a mathematics computing lab with the calculus instructor and teaching assistant two days a week. On the fifth day, students met in a classroom with both the physics and calculus instructors. The combined classes highlighted the connections between the calculus and physics concepts and challenged the students to extend the connections they had made during the week to more advanced problems. This program also included six hours of in-class time for the professors. Some instructors might find this time commitment too significant; however, advanced teaching assistants could ease some of the in-class burden on professors. Our instructors also met weekly to coordinate lessons and collectively prepare for classes.

About half the students in our class were enrolled in the honors program at the University of New Hampshire because the program staff believed that this course was appropriate for their students. The other students in our program were predominantly physics, mathematics, or electrical engineering majors because these students are advised to take physics and calculus the first semester of their first year of college. Other engineering majors took calculus during their first semester and physics during their second semester. The fact that our students were not selected randomly for this course has implications for our evaluation of the program, discussed in the "evaluation" section.

## PROGRAM IMPLEMENTATION

THE COMBINED course for calculus and physics has been offered for two years —one section in year 1 and expanded to two sections in year 2 because of students' interest and the support of the college dean and several department chairs. We discuss students' attitudes in the "evaluation" section; clearly these attitudes had an immediate effect on the growth and acceptance of the course in the college. Room limitations capped class size at twenty-four students; however, we plan to expand the course to thirty-six students in the future.

Although the students' enthusiasm got the program off to a good start, successful implementation of the curriculum depended on the instructors' paying close attention to how the students worked within the class and making appropriate changes along the way. To help in this ongoing assessment, we kept logbooks of the problems and successes of each class meeting and held collective debriefing sessions at the end of each semester.

The implementation of the student projects shows one aspect of how we implemented and changed the course over the two years. The student projects are longer and more complex context-rich problems than those worked on in class. In the first year, we assigned the students to work on one problem a month and gave them two weeks to complete it. The students worked in groups of three or four and met with one or both professors at least twice for each project. They presented their results either orally or in writing. The projects turned out to be quite difficult for the students, and, as a result, the students had to rely too heavily on help from the instructors. We concluded that this assignment was not helping the students become independent problem solvers.

The second year we made several changes to the student projects. First, we waited until the second semester to assign the projects so that the students would have had some experience with the four problem-solving steps. Second, we allowed the students to pick their own topics so that they would be working on something of interest to them. Finally, we made the main goal of the project the *process* of problem solving rather than the actual solution of the problem. This goal kept us from stepping in too early or often to help and gave us the leeway to let the students run into difficulties and solve them on their own. We believed that we were much more successful with this approach, although it was more challenging for us as instructors and required us to relinquish the problem selection to the students.

In one project that we considered a success, the group of students initially decided to find out how deep the dimples should be on a golf ball. The students quickly realized the extreme difficulty of this problem, but by using data from the literature, they were able to calculate the trajectories of an "ideal" golf ball (i.e., one with no spin or lift) and that of a real golf ball and to see a significant difference between them. The experience encouraged the students to use the literature; to revisit the ideas of force, velocity, and acceleration that were still troublesome to them; to understand the lift force both conceptually and numerically even though this topic had not been covered in class; and to use a computer to solve the differential equation.

# PROGRAM EVALUATION

## Overview

WE COLLECTED data over a two-year period using two cohorts of students—the students from our combined calculus and physics courses as well as the students enrolled in the regular sections of both calculus and physics concurrently. In the first year of the program (1998–1999), the cohort contained fewer than twenty-five students, so the results were not statistically significant. However, we used the results to guide the data collection for the following year. In the second year of the program (1999–2000), the cohort included about one hundred students because we offered two sections of the combined course and recruited volunteers more aggressively from the regular courses.

We collected data in three main areas—concepts and operations, attitudes and beliefs, and problem-solving ability. In addition, we collected a variety of data about the students' backgrounds to account for differences between the two groups. This background data included the students' SAT scores, their scores on a mathematics pretest, and the number of mathematics and physics classes each student had completed in high school.

## Concepts and Operations Data

For the area of operations, we compared the performance of students from both cohorts on solving standard problems. We wanted to find out whether the students in the combined course were able to carry out basic operations as skillfully as other students. To make this comparison, we designed and shared test questions with the instructors of the regular calculus and physics classes. Then we gave the examination problems to the students in both the combined and regular classes. We gathered the data and analyzed covariance (ANCOVA) to detect statistically significant differences between the performances of the two groups. On five out of six standard physics questions, the students in the combined course did as well as, or better than, the students in regular courses; on all the calculus problems, our students did as well as, or better than, the regular students.

To gauge conceptual gains in mechanics, we administered the Force Concept Inventory (FCI) (Halloun and Hestenes 1985), which measures students' ability to apply Newton's laws of motion to common situations. We administered this test before and after instruction, and the results were reported in terms of possible gain $h$ (Hake 1998), where $h$ is defined as follows:

$$h = \frac{(\text{class posttest average} - \text{class pretest average})}{(100 - \text{class pretest average})}$$

Hake collected data from across the country ($n = 6542$) and found that the gain for students from traditional lecture classes is $0.23 \pm .04$, whereas the gain for students from interactive engagement classes is $0.48 \pm .14$. The combined calculus and physics students gained 0.43 on the FCI, a performance that is consistent with the national average for interactive engagement classes.

## Attitudes and Beliefs

To ascertain students' attitudes and beliefs about mathematics and physics, we administered the Maryland Physics Expectation Survey (Redish, Saul, and Steinberg 1998). This test compares the students' answers with those of experts—physics instructors. The students used a Likert scale of agreement to respond to such statements as " Physical laws have little relation to what I experience in the real world." The questions fell into six categories: mathematics, independence, coherence, physics concepts, reality, and effort. This test was administered at the beginning and end of the academic year, and the results are quoted as movement toward or away from the expert view.

Our results were very similar to results from Dickinson College, which offers "Workshop Physics," a course that uses guided inquiry rather than lectures to help students construct their own knowledge (Laws 1991). Both the students from the combined course and the Dickinson College students moved *away* from the expert opinion in the categories of mathematics, independence, reality, and effort but moved *toward* expert opinion in coherence and physics concepts. We were surprised by the results in the mathematics category, especially because they did not agree with results from the clinical interviews that we had conducted on students' attitudes nor with written feedback. For example, in their feedback over the year, the students had shifted their view on learning from merely memorizing a set of facts for reiteration on tests to understanding concepts and ideas. This shift in their perspective is illustrated by the following comments:

- Before, I didn't really understand it [mathematics] and just sort of learned enough to get by, and then I'd forget the stuff after tests. Now I try to understand everything and for some reason, I don't forget it so easily. —*Student from the 1998–99 combined course.*
- The class stressed learning the main ideas and deriving the equations. This is much more useful than rote memorization of the equations. —*Student from the 1999–2000 combined course.*
- I realized how mathematics applies to real-life applications. I never realized the connection between calculus and the real world. —*Student from the 1999–2000 combined course.*

## Problem Solving

We also monitored students' progress in problem solving as they advanced through the combined calculus and physics program. We conducted several clinical interviews with our students and with students in the regular physics course to assess problem-solving skills. We had hoped to characterize similarities and differences between students who received direct problem-solving instruction in the combined program and students who did not receive problem-solving instruction in the calculus-based physics class.

For the interviews, we presented six students from each group with an array of problems and followed a think-aloud protocol. We transcribed the interviews and developed a coding scheme based on the gather, organize, analyze, and learn (GOAL) format (Serway and Beichner 2000) for problem solving, which is a version of the expert problem-solving heuristic documented by Pólya (1957).

One of the most striking differences between the two groups of students involved the students' uses of concepts and equations to solve problems. Five out of six students from the combined course talked about physics concepts or physics theory before beginning to solve a problem. In contrast, five out of six students from the regular physics classes focused their discussion on formulas and equations. For example, we noticed that the regular physics students tended to talk about the benefits of an equation list. These students claimed that if they had a list of equations in front of them, then they could "match up" the equation to solve the given problem.

Furthermore, in their open-ended feedback, many students from the combined courses asserted that they felt more confident and comfortable as problem-solvers. They believed that their problem solving skills had improved throughout the year and said that they were more willing to spend time working out problems at the end of the course than they had been in the beginning. Some examples of their feedback include the following:

- I feel more confident that, given enough time, I can solve most problems. I understand calculus much more than when it was presented in high school, because in high school, we learned how to solve problems, not why we were doing what we were doing. —*Student from the 1998–99 combined course.*
- I learned a new way to look at calculus and physics. And more importantly, I have a much better understanding of how to tackle hard problems. —*Student from the 1999–2000 combined course.*

## CONCLUSION

THE DATA that we gathered on students' attitudes coupled with the enthusiasm of the instructors indicate that the combined calculus and physics course for first-year college students was a success. Not only did our students appreciate the value of mathematics and physics more than they had in the past, they also understood mathematics and physics concepts at a deeper level. The students embraced the integrated approach to learning these two subjects and began building a foundation of connected knowledge. These students have expanded and grown more confident in their problem-solving abilities. They were able to solve standard problems as well as, or better than, students in the traditional physics and calculus classes. Furthermore, because the students in our course approached the subjects more deeply, we as instructors were more enthusiastic about teaching. The small class size allowed for more personal interactions among students and between students and professors; this interaction is an essential part of the learning process. We shared the students' positive attitudes with other professors and the dean in the hope that the favorable comments would help ensure the continuation of the course.

We overcame the practical difficulties in implementing an interdisciplinary course by cooperating closely, communicating, and agreeing to similar goals. Without common goals to unite our efforts, the development of this course would have been a difficult process. Researchers must continue to explore the effects of such integrated approaches on students' understanding of, and attitudes toward, mathematics. Implementers must continuously evaluate and refine integrated programs to successfully foster connections among the disciplines and enhance students' development of problem-solving and lifelong-learning skills.

## REFERENCES

Arcavi, Abraham, Cathy Kessel, Luciano Meira, and John P. Smith III. "Teaching Mathematical Problem Solving: An Analysis of an Emergent Classroom Community." In *CBMS Issues in Mathematics Education, vol. 7:* Research in Collegiate Mathematics Education III, edited by Alan H. Schoenfeld, James Kaput, and Ed Dubinsky, pp. 1–70. Providence, R.I.: American Mathematics Society, 1998.

Brown, Ann L. "Analogical Learning and Transfer: What Develops?" In *Similarity and Analogical Reasoning*, edited by Stella Vosniadou and Andrew Ortony, pp. 369–412. Cambridge, England: Cambridge University Press, 1989.

Glaser, Robert. "Expert Knowledge and Processes of Thinking." In *Enhancing Thinking Skills in the Sciences and Mathematics*, edited by Diane F. Halpern, pp. 63–75. Hillsdale, N.J.: Lawrence Erlbaum Associates, 1992.

Greeno, James G., and the Middle-School Mathematics through Applications Project Group, "Participation as Fundamental in Learning Mathematics." In *Proceedings of the Nineteenth Annual Meeting of the North American Chapter of the International Group for the Psychology of Mathematics Education*, edited by John A. Dossey, Jane O. Swafford, Marilyn Parmantie, and Anne E. Dossey, vol. 1., pp. 1–14. Bloomington–Normal, Ill.: University of Illinois, 1997.

Hake, Richard R. "Interactive-Engagement versus Traditional Methods: A Six-Thousand-Student Survey of Mechanics Test Data for Introductory Physics Courses." *American Journal of Physics* 66 (January 1998): 64–74.

Halloun, Ibrahim Abou, and David Hestenes. "The Initial Knowledge State of College Physics Students." *American Journal of Physics* 53 (November 1985): 1043–55.

Heller, Patricia, Ronald Keith, and Scott Anderson. "Teaching Problem Solving through Cooperative Grouping; Part 1: Group versus Individual Problem Solving." *American Journal of Physics* 60 (July 1992): 627–36.

Larkin, Jill H. "Information Processing Models in Science Instruction." In *Cognitive Process Instruction*, edited by Jack Lochhead and John J. Clement, pp. 109–18. Hillsdale, N.J.: Lawrence Erlbaum Associates, 1979.

Laws, Priscilla. "Calculus-Based Physics without Lectures." *Physics Today* 24 (December 1991): 24–31.

Leonard, William J., and William J. Gerace. "Students' Reflections on an Introductory Physics Course." University of Massachusetts Physics Education Research Group Technical Report, University of Massachusetts, August 1994.

Lester, Frank K., and Joe Garofalo. "Metacognitive Aspects of Elementary School Students' Performance on Arithmetic Tasks." Paper presented at the annual meeting of the American Educational Research Association, New York, March 1982.

National Research Council (NRC) Center for Science, Mathematics, and Engineering Education. From Analysis to Action: *Undergraduate Education in Science, Mathematics, Engineering, and Technology.* Washington, D.C.: National Academy Press, 1996.

National Science Foundation (NSF) Directorate for Education and Human Resources. *Project Impact: Disseminating Innovation in Undergraduate Education Conference Proceedings.* Arlington, Va.: National Science Foundation, 1994.

———*Shaping the Future: New Expectations for Undergraduate Education in Science, Mathematics, Engineering, and Technology.* NSF Report No. 96–139. Arlington, Va.: National Science Foundation, 1996.

Pólya, George. *How to Solve It: A New Aspect of Mathematical Method.* 2nd ed. Princeton, N.J.: Princeton University Press, 1957.

Redish, Edward, Jeff Saul, and Richard Steinberg. "Student Expectations in Introductory Physics." *American Journal of Physics* 66 (July 1998): 212–24.

Schoenfeld, Alan H. "Beyond the Purely Cognitive: Belief Systems, Social Cognitions, and Metacognitions as Driving Forces in Intellectual Performance." *Cognitive Science* 7 (October–December 1983): 329–63.

———. *Mathematical Problem Solving.* Orlando, Fla.: Academic Press, 1985.

Serway, Raymond A., and Robert J. Beichner. *Physics for Scientists and Engineers.* Fort Worth: Saunders College Publishing, 2000.

Sherin, Bruce L. "How Students Understand Physics Equations." *Cognition and Instruction* 19 (December 2001): 479–541.

Wilson, Jack M. "The CUPLE Physics Studio." *The Physics Teacher* 32 (December 1994): 518–23.

9

# Integrating Mathematics and Science in a Preservice, Elementary-Level Teacher Education Program

Cengiz Alacaci
George E. O'Brien
Scott P. Lewis
Zhonghong Jiang

MAKING CONNECTIONS between school mathematics and science has long been viewed favorably among educators. Various forms of integration exist, and a multitude of curriculum-integration models have been developed (Berlin and White 1995). Teaching mathematics in relation to science supports students' learning by providing a meaningful context in which they can see the applications of abstract concepts. Furthermore, integration helps instill positive attitudes in children by showing them the utility of mathematics. From the perspective of science education, mathematics offers the tools for quantifying, representing, and analyzing scientific phenomena and fosters a sense of objectivity in science (Lonning and DeFranco 1994).

These benefits, however, are only a small part of what integrating mathematics and science has to offer. The landscape of integration at the elementary school level is colorful and multifaceted. The real power of integration lies in the other, often less-well-known facets of integration, including integration in the processes of inquiry and methodology and of teaching and learning. In this chapter, we highlight the existing theoretical models of integration. Then we present and discuss our experiences with integrating mathematics and science in an elementary-level teacher education program at Florida International University (FIU).

# WHY SHOULD MATHEMATICS AND SCIENCE BE CONNECTED?

CURRENT REFORM documents in mathematics and science education strongly endorse connecting the two disciplines. In *Principles and Standards for School Mathematics* (NCTM 2000), one of the five Process Standards common to all grade bands is Connections. The Connections Standard recommends fostering mathematical experiences in which students work on problems arising from other disciplines, particularly science. Furthermore, the document suggests that "the link between mathematics and science is not only through content but also through process" (p. 66). The processes of scientific inquiry can inspire and support the processes of mathematical problem solving. The authors of *Principles and Standards* clearly see the benefits that connecting mathematics and science offers for developing process skills common to both disciplines.

Similarly, *National Science Education Standards* (National Research Council 1996) is quite explicit about the value of linking school mathematics and science. One of the science education program standards is devoted to coordinating and linking science and mathematics programs. The document notes that "coordination of science and mathematics programs provides an opportunity to advance instruction in science beyond the purely descriptive. Students gathering data in a science investigation should use tools of data analysis to organize these data and to formulate hypotheses for further testing" (pp. 218–19).

# ISSUES AND MODELS OF INTEGRATION

INTEGRATING MATHEMATICS and science has been of interest among educators and curriculum developers for a long time (McNeil 1985). A number of terms, such as *connecting, correlating, integrating, coordinating*, and *interacting*, are used to describe and differentiate among the different ways of linking the disciplines (Berlin and White 1995). Below we describe two models that frame the issues surrounding integration.

## The Davison, Miller, and Metheny model

Davison, Miller, and Metheny (1995) identified the following five ways of integrating mathematics and science:

1. *Discipline-specific integration* refers to integrating different branches or strands of mathematics, for example, number, geometry, and statistics. In this model, different

branches of science, such as chemistry and biology, can be integrated similarly within the subject of environmental science.

2. *Content-specific integration* concerns coordinated teaching of one learning objective from mathematics and one from science, for example, when proportions—a mathematical objective—is taught in the context of simple machines—a science objective.

3. *Thematic-integration* is used to connect mathematical concepts and skills with a scientific context or a theme. For instance, when students are exploring such environmental science topics as oil spills, they can learn concurrently about such measurement concepts as volume and area.

4. *Process integration* refers to making explicit connections between the process skills of science and mathematics. Important scientific process skills include observing, communicating, classifying, predicting, inferring, controlling variables, analyzing data, testing hypothesis, defining operationally, and experimenting. Important mathematical process skills include problem solving, reasoning, communicating, representing, and connecting (NCTM 2000). Teaching these process skills can span more than one class period and remains an overall goal of instruction in these disciplines. Obvious similarities exist between some of these skills (e.g., communication), and more implicit yet robust parallels exist among the others. For example, representation skills in mathematics require facility with graphical, tabular, numerical, geometric, algebraic, written, and verbal representations, skills that parallel the data analysis component of the scientific-inquiry process.

5. *Methodological integration* links more general skills, for example, problem solving in mathematics and inquiry and discovery in science.

Although the Davison, Miller, and Metheny (1995) model offers powerful ways of understanding *what to integrate* when we connect mathematics and science, it does not directly address the relative primacy of the two disciplines in integration. Realistically, the undertaking of creating and implementing a fully integrated mathematics and science curriculum that treats the two disciplines equally may not be feasible.

## The Lonning and DeFranco model

Addressing the potential richness of mutual connections, Lonning and DeFranco (1997) present a continuum model of integration with regard to the relative primacy of mathematics and science. According to their work, the continuum has the following five components:

1. *Independent mathematics*, in which mathematics concepts are taught separately with no explicit connections to science
2. *Independent science*, in which science concepts are taught separately with no explicit connections to mathematics
3. *Integration with a mathematics focus*, in which teaching mathematical concepts is the primary focus and science is taught in support of the mathematical concepts. When the integrated-science concepts belong to a lower-level curriculum than the mathematical concepts or when science simply establishes a familiar context for the mathematical concept or procedure, integration has a mathematics focus.
4. *Integration with a science focus*, in which science concepts and activities are of primary importance, whereas the mathematical concepts come from a lower grade level to support the science concepts
5. *Balanced mathematics and science integration*, in which the goals of mathematics instruction are on equal footing in importance with the goals of science instruction and in which objectives for each discipline come from the same grade level

In describing what knowledge and skills can be integrated (Davison model) and how to prioritize integration (Lonning model), both these models approach integration from a curriculum perspective. However, other important integration approaches exist for mathematics and science.

Berlin and White (1995) suggest that integration should also be considered from the perspectives of learning and teaching. Relative to learning, coherent integration of mathematics and science should be based on similar theories. Take, for instance, constructivism, which many mathematics and science educators value as a tool. They consider the following principles of constructivism essential:

- Knowledge is best attained when the learner is actively involved in the process.
- Reflecting on the physical and mental actions aids consolidation of the knowledge.

- Social interaction among students and between students and the teacher facilitates learning. (Reys et al. 2001)

Because recent reforms in both mathematics and science emphasize process skills that rely on these constructivist principles, constructivism can be an important pedagogical bridge linking mathematics and science.

Regarding teaching, congruent methods can support the goals of both science and mathematics. Both disciplines rely on instruction that furnishes challenging and worthwhile tasks, facilitates rich communication, and offers opportunities for cooperative work. Furthermore, congruent teaching of mathematics and science should give students opportunities to use hands-on materials and laboratory instruments. Assessment techniques should include portfolios, projects, and performance tasks as well as paper-and-pencil tasks and multiple-choice tests (NCTM 2000; NRC 1996).

Below we describe our efforts to connect mathematics and science in a preparation program for elementary-level preservice teachers. We also describe the ways in which we integrated the disciplines and analyze how our model compares with the theoretical framework outlined above.

## CONNECTING MATHEMATICS AND SCIENCE IN AN ELEMENTARY-LEVEL TEACHER EDUCATION PROGRAM

FLORIDA INTERNATIONAL University (FIU) is located in Miami-Dade County and is the largest producer of new teachers in the South Florida region. The elementary-level education program has been revised recently to meet the needs of students better. One change requires the program to make explicit connections among "disciplines that structure the field of elementary education, such as reading, language arts, mathematics, science, social studies, art, health and physical education, and music" (Florida International University 1998). With this vision in mind, the faculty has sought ways to make explicit connections between science and mathematics methods courses that had been previously taught as isolated units. Mathematics and science methods courses are requisites for college juniors in the elementary education program at FIU.

### The science methods course

The science methods course features a project-based science (PBS) approach that follows the Krajcik, Czerniak, and Berger (1999) model. In this approach, students are actively engaged in developing a semester-long science project based on a "driving question." The driving question must meet such criteria as feasibility, worth, contextualization, meaningfulness, and sustainability. To stimulate the process, the instructor poses a central question for the class, for example, "What is in our water?" Working in cooperative groups, students determine their own questions for independent group projects related to the driving question. In a recent class section, the students came up with such questions as "What type of water is safe to drink?" and "How do water-treatment methods affect the quality of our drinking water?" Students then read about the topic and gather information to guide their thinking about the essential variables related to the driving question. As the project develops, they create an "investigative question" that is more specific and operationally definable. Sample investigative questions are "What are the levels of chlorine, pH, and phosphate found in bottled, tap, and filtered water?" and "Which water filter is most effective in removing chlorine from our drinking water?"

To answer these questions, students either conduct their own investigations and measurements in the science lab or collect data from such other resources as local institutions and Web sites. Once students obtain their data set, they typically transform the data into a graph or a table. Students often use such mathematical calculations as percents, ratios, mean, median, and mode to analyze the data. These mathematical tools allow students to capture patterns to help them answer their investigative questions.

At the beginning of the semester, the students receive the set of project guidelines shown in figure 9.1. These guidelines explicitly ask the students to reflect on the mathematical tools that they use in the projects. For example, for graphing, they must consider whether the title that they chose is descriptive, whether they should use a legend, whether the axes are properly labeled, whether the numerical scales are appropriate, and why they think that their representation is an effective way to present the data. They are also asked to reflect on how the chosen mathematical tools, such as percents, fractions, mean, median, and mode, show the important features of the data. The purpose is to raise students' awareness of not only how to use a mathematical tool but also when to use it.

The mathematics methods professor visits the students while they work on their projects in the science methods class and answers their questions about the mathematics components. Topics frequently raised during those visits include when to use a pie chart (to compare parts in relation to a whole), when to use a bar graph (to compare categorical data), when to use line graphs (when the data are in ordinal scale),

## Science Project Guidelines

A. Provide a concept map about the topic, including at least eight concepts and linking words.

B. List the investigative question.

C. Attach a copy of the data set used.

D. Provide the source of the data set (encyclopedia, Internet, student-gathered, etc.).

E. Represent and analyze the data.

*Note that this section will connect with the mathematics methods course and will be reviewed by the mathematics methods professor.*

1. *Is the choice of data-representation tool (e.g., pie chart, bar graph, column graph, line graph, scatterplot, stem-and-leaf graph, box plot) appropriate?*
2. *Is an explanation of why this tool is used given?*
3. *Does the data representation reflect the following?*
   - *A match between the nature of the data and the title*
   - *The use of legends*
   - *The identification of the axes*
   - *A choice of scale in the numerical axes*
4. *If data statistics (e.g., mean, median, mode, range, variance, and standard deviation) are used, explain how the choice of statistics facilitates showing the pertinent features of the data.*

F. Create a set of findings or conclusions based on answering the investigation question, "Do the data support it?"

G. Consider additional research questions.

H. Discuss the social ramifications of the investigative question and findings.

I. Make connections with the NRC Science Education Standards.

J. Make connections with the College of Education goals.

Fig. 9.1. Guidelines for reporting science projects in the science methods course, with the mathematics component highlighted in italics. Adapted from www.figurethis.org (Figure This Math Challenges for Families, 2002).

and how to compute averages when the data include outliers (use median or mode instead of arithmetic mean). Students often use such technological tools as spreadsheets to create graphs.

The use of mathematical data analysis and representation tools in science projects is an instance of content-specific integration from the Davison, Miller, and Metheny (1995) model. In this scenario, students learn or relearn and apply mathematical knowledge pertaining to graphing, averages, percents, and ratios for the purpose of scientific inquiry. The primary purpose of their work is to conduct the scientific project and to answer the investigative question, therefore the project also has a science focus within Lonning and DeFranco's (1997) framework.

## The mathematics methods course

This course helps preservice teachers experience the vision of mathematics education reform as outlined in the NCTM's *Standards* documents (NCTM 1991, 2000). Therefore, modeling teaching through problem solving is a recurring theme as students pursue the twin goals of (*a*) revisiting the content knowledge of elementary-level mathematics and (*b*) addressing the related pedagogical issues. Each class session covers teaching issues pertaining to a particular content strand, for instance, number concepts, computation, measurement, or statistics. The class time is divided into three activities: weekly mathematical investigations of the selected content strands, instructional simulations, and in-depth discussions that link the day's activities and students' assigned readings for the week. The investigations demonstrate ways to teach mathematical concepts during problem solving as well as ways in which mathematics and science can be integrated in the elementary-level curriculum.

Prior to class, the preservice teachers work at home on a mathematical task about the coming week's topic. During class, the students share their written solutions with a group of two other classmates and compare their methods and answers. Then the professor asks several individuals to share their solutions with the whole class and notes the similarities and differences in the strategies. After all the different solution methods are considered, the students reflect on the educational aspects of their experience. For example, they consider the mathematical concepts and processes involved in the task, the power of the group work, and the role of technology in problem solving.

Some weekly investigation tasks are embedded in scientific contexts and demonstrate thematic integration, which is relatively better known than the other types of integration. For example, for one investigation, we adapted a graphing activity in an environmental science context from the National Center for Ecological Analysis and Synthesis (1997). In this activity, students are given quantitative information about trash collected from a beach and are asked to represent the data in a graph (e.g., pie chart or line or bar graph). They are also asked to explain their choice of graph and compare it with, and relate it to, their peers' choices. In this task, the purpose is to discuss the choice of graphs to answer particular questions.

Process and methodological integration, an often-overlooked type of integration, can also be modeled effectively. One example of this type is the weekly mathematical investigation on the Popcorn Containers Task in figure 9.2. The Popcorn Containers Task has a counterintuitive answer. Because all the shapes are made from paper of the same size, many students predict that the shapes will have the same volume or hold the same amount of popcorn. The task lends itself to multiple methods of solution. Students may use an empirical approach by first making the shapes, then filling the containers with materials like popcorn, marshmallows, cotton balls, and so forth, and counting and comparing the contents. Or, they may use an algebraic approach by computing the volumes of each shape

Popcorn Containers

Imagine that you want to make a popcorn container by folding and taping together an 8.5-by-11-inch (that is, regular paper size) piece of rectangular cardboard. It is possible to make at least four different shapes—a short cylinder, a tall cylinder, a short rectangular prism with a square base, and a tall rectangular prism with a square base, as seen below. You may assume that you do not need to cover the bottoms and tops of the containers.

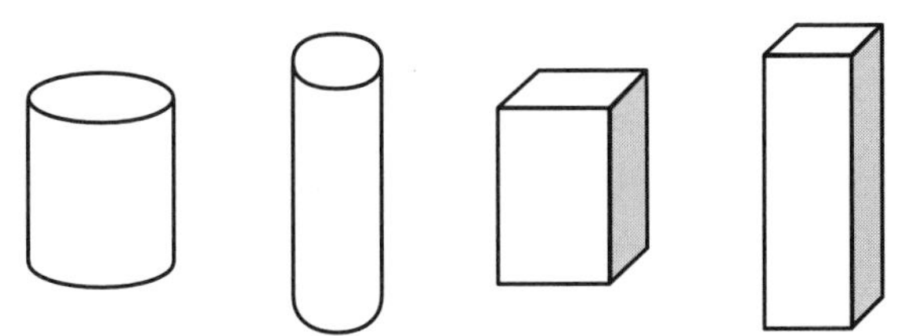

Do these shapes hold the same amount of popcorn? If not, which one holds the most? How do the different shapes compare with regard to the amount of popcorn that they can hold? Please give evidence to support your answer, and be sure to show your work.

Fig 9.2. The Popcorn Containers task

using volume formulas, a multistep process that first requires the students to find the radius from the circumference (8.5 inches for a tall cylinder or 11 inches for a short cylinder). Although the Popcorn Containers Task is not embedded in a scientific context, the solution process resembles scientific inquiry in the following ways:

1. It starts with a question.
2. It develops a hypothesis (that the shapes hold the same amount).
3. It devise a plausible empirical or algebraic plan to answer the question.
4. It identifies variables.
5. It develops a model of relationships among the variables, such as how the independent variables of height, width, and radius relate to the dependent variable of volume.
6. It collects the data either by computing the volumes of the shapes or by filling them with a substance and comparing the contents.
7. It analyzes and interprets the results in relation to the original question.
8. It extends the new insights by posing such questions as "Why does radius affect the volume more than the length?" or by using such two-dimensional shapes as rectangles with the same perimeter to consider the question "Which one holds the most space or area?"

After discussing different solutions and strategies in class, the preservice teachers reflect on the mathematical process. Issues frequently raised include how they felt dealing with the problem, what the crucial junctures were in the process, what role technology (e.g., calculators) played in the solution, and how the experience resembled the project-based science (PBS) process.

The students' reflection also includes connecting the assigned readings of the week with their experience in the investigations, for example, the Popcorn Containers task. Using a variety of additional tools, such as case studies, vignettes, and video demonstrations showcasing inquiry-based lessons in mathematics, the preservice teachers discuss the dynamics of classrooms featuring active-inquiry learning. Their experience in PBS often helps them draw powerful analogies between mathematics and science in these discussions, which also include reflection on the topic of integrated teaching.

Using such tasks as the Popcorn Containers in our mathematics methods class allows us to model process and methodological integration, a type of integration that is suggested in the NCTM's *Principles and Standards for School Mathematics* (2000) and included in the Davison, Miller, and Metheny (1995) model. As they experience this integration model, the preservice teachers can see why educators in both disciplines value active construction of knowledge as a way of teaching and learning. Such construction takes place through inquiry in science education and problem solving in mathematics. Furthermore, the preservice teachers can see the striking inquiry-related similarities that exist between the two disciplines. Reflection on these similarities can benefit learning in the two disciplines and help new teachers construct a classroom culture in which learning by active inquiry—rather than by passive receipt of knowledge—is the norm.

Integration of mathematics and science in our model at FIU is further strengthened by the similarities in the methods of teaching and in the implicit theories of student learning. Preservice teachers work cooperatively on challenging and worthwhile tasks using technology. They extend their knowledge and skills by solving problems for which they do not know the answer, and they use active and rich communication with their peers and the instructor. The role of the instructor is analogous to that of a coach, and the students do the work by being actively engaged with a task. The weekly mathematical investigation illustrates integration in the process and methodology of inquiry of the two disciplines. We believe that the authors of *Principles and Standards* (NCTM 2000) had this type of integration in mind when they stated, "The link between mathematics and science is not only through content but also through process" (p. 66).

## CONCLUSION

THE IDEAS that we have presented illustrate the complex issues involved in integrating mathematics and science at the preservice-teaching level. Integration offers more than just science contexts for mathematical concepts or mathematics tools for scientific exploration. As we demonstrated in this chapter with the discussions of modes of inquiry and of knowledge production and acquisition, the connection runs deeper. Therefore, when we integrate mathematics and science, we must consider the congruence in the methods and processes of teaching and assume that the two disciplines share similar theories of students' learning. Furthermore, we must note the availability of many alternatives to full-scale integration, including content-specific integration, thematic integration, and process and methodological integration at multiple points in the curriculum. Content-specific and thematic connections may enhance the learning of one subject more than the other at given times, whereas process and methodological integration will yield long-term benefits for *both* mathematics and science.

So that preservice teachers can reap the benefits of integration when they actually start teaching their own classes, they must experience viable models of integration in their methods courses. Furthermore, they need active support from the teaching faculty to ensure that they develop the necessary pedagogical knowledge.

As faculty, we started our collaboration with the desire to leave our "isolated boxes" of single-subject courses and build connections between mathematics and science for the benefit of the preservice professionals who will be teaching both disciplines. With a common vision, we started visiting one another's classes, sharing insights, and offering suggestions that would build continuities between the courses. We met regularly to discuss action research projects supporting integration and developed a collaborative research agenda. We attended and gave presentations at one another's professional conferences and coauthored manuscripts. For us, the integration—as powerful as it is on its own—seems to be an open-ended process through which we continuously discover new ideas and fine-tune the old ones. For our next steps, we are evaluating how integrating mathematics and science in preservice teaching courses affects our graduates' teaching practices.

## REFERENCES

Berlin, Donna F., and Arthur. L. White. "Connecting School Science and Mathematics." In *Connecting Mathematics Across the Curriculum,* 1995 Yearbook of the National Council of Teachers of Mathematics (NCTM), edited by Peggy A. House and Arthur F. Coxford, pp. 22–33. Reston, Va.: NCTM, 1995.

Davison, David M., Ken W. Miller, and Dixie L. Metheny. "What Does Integration of Science and Mathematics Really Mean?" *School Science and Mathematics* 95 (May 1995): 226–30.

Figure This Math Challenges for Families. "Math Challenge #3: Popcorn." www.figurethis.org/challenges/c03/challenge.htm (13 January 2003).

Florida International University. "Elementary Education Program Outcomes." Miami: Faculty of the College of Education, Florida International University, 1998. Photocopy.

Krajcik, Joseph, Charlene M. Czerniak, and Carl F. Berger. *Teaching Children Science: A Project-Based Approach.* Boston: McGraw-Hill College, 1999.

Lonning, Robert A., and Thomas C. DeFranco. "Development and Implementation of an Integrated Mathematics/Science Preservice Elementary Methods Course." *School Science and Mathematics* 94 (January 1994): 18–25.

———. "Integration of Science and Mathematics: A Theoretical Model." *School Science and Mathematics* 97 (April 1997): pp. 212–15.

McNeil, John D. *Curriculum: A Comprehensive Introduction.* 3rd ed. Boston: Little, Brown, & Co., 1985.

National Center for Ecological Analysis and Synthesis (NCEAS). *Kids Do Ecology.* 1997. www.nceas.ucsb.edu/nceas-web/kids (13 January 2003).

National Council of Teachers of Mathematics (NCTM). *Professional Standards for Teaching Mathematics.* Reston, Va.: NCTM, 1995.

———. *Principles and Standards for School Mathematics.* Reston, Va.: NCTM, 2000.

National Research Council. *National Science Education Standards.* Washington, D.C.: National Academy Press, 1996.

Reys, Robert E., Mary M. Lindquist, Diana V. Lambdin, Nancy L. Smith, and Marilyn N. Suydam. *Helping Children Learn Mathematics.* 6th ed. New York: John Wiley & Sons, 2001.

# 10

# Integrated Mathematics through Mathematical Modeling

Heidi L. Keck
Johnny W. Lott

INTEGRATED MATHEMATICS can be studied through a mathematical-model lens. To study integrated mathematics this way, however, teachers need to (*a*) know how and why mathematical models are derived, (*b*) create their own models, and (*c*) think about the relationship between the models and the mathematics that is being integrated. This chapter describes one curricular experiment—implemented at the University of Montana in Missoula—that has enabled prospective teachers to see the power of mathematical modeling in integrating mathematics.

The experimental course was based on an understanding of integrated mathematics as "a holistic mathematical curriculum that consists of topics chosen from a wide variety of mathematical fields and blends those topics to emphasize the connections and unity among fields" (Beal et al. 1990). In addition, we considered integrated mathematics to be a curriculum that emphasizes the relationships among topics within mathematics as well as between mathematics and other disciplines.

The prospective teachers in our course were preparing to teach a curriculum called *Integrated Mathematics: A Modeling Approach with Technology* (MCTM/SIMMS 1998). This integrated-mathematics curriculum is (*a*) integrated and interdisciplinary, (*b*) problem-centered and applications-based, (*c*) reliant on the appropriate use of technology as a learning and teaching tool, and (*d*) dependent on multiple delivery modes to accommodate the multiple learning styles of students (Burke and Lott

1993). The curriculum also incorporates the conceptual mathematization of de Lange (1989). Finally, it allows learners to develop concepts through experiencing situations, or models, that can be manipulated, organized, and structured according to mathematical aspects.

Mathematical modeling is basic to this curriculum and demands that students reduce problem situations to fundamental mathematics and understand the processes involved (Burke and Lott 1993). Research on how to assist preservice teachers in developing confidence in mathematical modeling, as well as in maintaining a classroom atmosphere conducive to problem solving, has identified several important instructional strategies for teacher educators (Camerlengo 1993; Holmes Group 1986; Lester 1985, 1994; Thompson 1985). These strategies include (*a*) allowing teachers to experience mathematical modeling or problem solving from the perspective of the problem solver, (*b*) encouraging them to reflect on their own thought processes, and (*c*) helping them to understand the interconnectedness of mathematics, both with other disciplines and within mathematics.

To prepare prospective teachers to use modeling and integrated mathematics in their classrooms, we designed a senior-level undergraduate class focusing on applications, or mathematical modeling, using technology. We chose applications from across mathematical fields and required our students to use the library or Internet for their research. The class was required for students pursuing a secondary-level mathematics education major, and its prerequisites included two semesters of calculus and an introductory course in abstract mathematics. We designed the modeling component of the class to teach students to study existing models, give an analytical appraisal of each model, and create mathematical models of their own.

To create their models, the students worked cooperatively on problems, used technology to investigate mathematical relationships, described models in written essays, and made oral presentations. In this way, they integrated mathematics with ideas from science, communications, and technology.

Alternative methods of instruction are a necessary part of teaching mathematics to reach the widest possible audience (Cuevas and Driscoll 1993), and researchers believe that teachers who have experienced a variety of instructional methods are better able to use a variety of instructional methods themselves (Castle and Aichele 1994). Because we incorporated various methods in our teaching of the course, our students became more comfortable teaching with these methods. The students not only learned modeling but also connected mathematics with other subjects and learned a variety of instructional techniques.

The class met in a computer lab for fifty minutes, three days a week. The students were encouraged to use calculators, computers, and software and used *Mathematical Modeling in the Secondary Curriculum* (Swetz and Hartzler 1991) and *Mathematical Modelling* (Berry and Houston 1995) as textbooks. The specific technology that they used included the following:

- Microsoft Word (word processing software)
- Microsoft Excel (spreadsheet software)
- TI–82 or TI–92 calculators (graphing calculators)
- The Geometer's Sketchpad or Cabri (geometric software)
- Data Desk (statistics software)
- TEMATH (graphing software)
- Netscape (Web browsing software)
- Mathematica or MathCad (symbolic manipulators)

Before starting their modeling assignments, the students completed work designed to acquaint them with the various technologies.

For the course, we assigned four modeling projects. The students worked in teams of two and had approximately two weeks to complete each project. In addition, the students completed one mathematical writing assignment and eleven assignments involving the use of the technology that was introduced during the semester. Each of these assignments was completed in about one week. Because of the timing of the assignments, the modeling projects overlapped the technology assignments.

The concepts involved in mathematical modeling were divided into two parts:

1. The first part included determining the problem, identifying assumptions, and creating the model.
2. The second part included obtaining information from the model, validating the model, analyzing the underlying structures, and generalizing the model, if possible.

We introduced the first part of the modeling process with in-class activities. To study the process of building models based on theoretical concerns, we adapted the following activity for second-year-algebra students from a problem by Swetz and Hartzler (1991).

## An Irrigation Problem

A linear irrigation system consists of a long water pipe on wheels with sprinklers mounted at regular intervals along the pipe. The system moves slowly across a rectangular field to give all parts of the field the same designated amount of water. The manufacturer wants you to decide how to space the sprinklers on the pipe to give the most uniform coverage possible. After you have established the best spacing, you should decide how fast the system should move across the field to drop one inch of water in one pass.

The specifications are as follows:

1. Each sprinkler head produces the spray pattern shown in figure 10.1. The flow is 5 gallons a minute for each sprinkler. Water falls uniformly on the area between two concentric circles with radii of 1 foot and 20 feet.

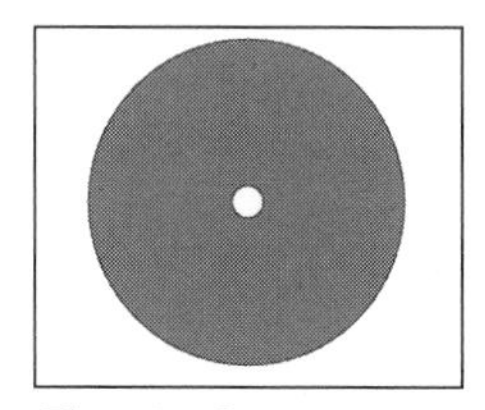

Fig. 10.1 Spray pattern

2. The field is 1000 feet wide and 2000 feet long.
3. To avoid runoff, spray patterns should overlap no more than twice.

The students worked in pairs during two class periods to define uniform coverage and decide how to measure coverage. This work corresponds to the components of part 1 of the modeling process—identifying the problem (e.g., what is uniform coverage?), making necessary assumptions, and creating a model (e.g., how is coverage measured?). The students struggled to define "uniform" as anything other than "exact equal amounts of water" on every section of the field and then came to the realization that defining "uniform" within the problem's specifications was impossible. This activity set the tone for the students' in-depth involvement in their four assigned projects.

At the end of the first week of class, the students chose their problems for project 1 from a provided list. We asked the students to use their chosen problem to set up a mathematical model *without* answering the question, validating the model, or providing any analysis of the model itself. Four groups in the class chose the problem "You are provided with a roll of toilet paper. Formulate a model to predict its length" (Berry and Houston 1995). All four of the paper-roll projects were written in three distinct sections: problem identification, assumptions, and model. All the papers listed explicit steps that one would use to predict the length of a particular roll. All four groups derived single equations to predict the length in terms of two or three parameters and to serve as the model. However, the students had less success with making and justifying assumptions. One paper did not attempt to justify the assumptions made. Two of the papers attempted to justify making the assumptions but only by stating that the assumptions were necessary for their particular approach. In these papers, although the reader might infer why the students made these assumptions, the students presented no actual evidence.

All the papers lacked transitional phrasing with regard to the mathematics. Instead of incorporating mathematical equations and variable definitions into sentences, the students placed the equations on separate lines. And rather than include facts in their paragraphs, the students merely listed them.

When we returned these papers to the students, we reminded them that the modeling projects were also communication or writing assignments. We stressed that sentences needed to be contained in paragraphs and that paragraphs needed a topic sentence and supporting arguments. We encouraged the students to refer to mathematics textbooks for examples of mathematical writing.

At this point, we introduced the second part of the modeling process—the processes of obtaining information and validating and generalizing the model. We assigned the students to work on a revised version of the irrigation problem for two class periods. This time, they used the following model presented by Swetz and Hartzler (1991, p. 52).

### The Irrigation Problem Revisited

One model sets the first sprinkler head at the origin and the second sprinkler head at the point $(d, 0)$. From this, a function, $C(x)$, was defined to give the time that a particular point is under the sprinkler as a function of its $x$-coordinate, as shown below.

$$C(x) = \begin{cases} 2(\sqrt{400 - x^2} - \sqrt{1 - x^2}) \text{ for } 0 \leq x \leq 1 \\ 2\sqrt{400 - x^2} \text{ for } 1 \leq x \leq (d - 20) \\ 2\sqrt{400 - x^2} + \sqrt{400 - (x - d)^2}) \text{ for } (d - 20) \leq x \leq \dfrac{d}{2} \end{cases}$$

Then, uniform coverage was defined to be that which has the least difference in the maximum and minimum coverages.

For $d = 30$, the graph of $C(x)$ is shown in figure 10.2. The horizontal axis represents the position along the width of the field (assuming that the sprinkler moves across the length of the field), and the vertical axis represents the time that the sprinkler will be over any place in the field. The best value of $d$ can be found by making a chart of values of the least time and greatest time difference for varying $d$ values.

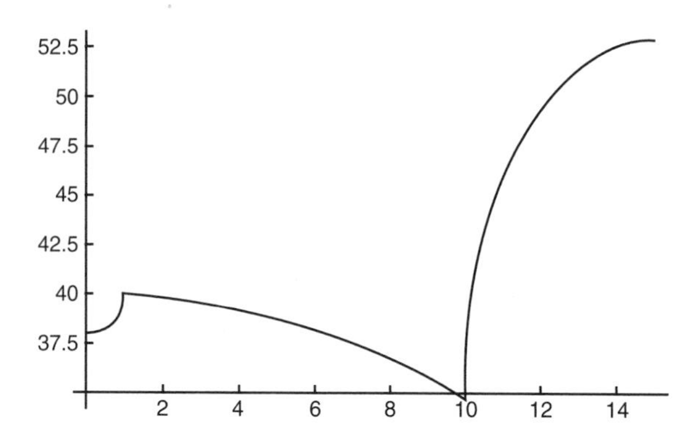


Fig. 10.2. Position on field versus time under sprinkler

On the basis of the foregoing model, the students determined the spacing of the sprinklers and the speed of the system as it moved across the field. They also validated the model and reflected on the assumptions. We encouraged the students to generalize the model to include different specifications. During the class discussion, the students focused on the following three assumptions: (1) a single sprinkler head producing uniform flow was unrealistic, (2) the water falling at the edges of the field was not being measured adequately, and (3) the lack of agreement on uniform coverage was problematic enough to make the model invalid. The students questioned the ability of the average farmer to reproduce their calculations. They overlooked the concept of "sufficiently accurate" as well as the fact that an engineer at an irrigation manufacturing company must solve problems similar to this one before an irrigation product can be marketed.

After two days of working on the revisited irrigation problem, the students received their second project assignment. The new project involved reading a previously developed model, using it to answer a specific question, and analyzing the model. This assignment continued the emphasis on the second half of our modeling process. The students chose from four models. Three were established models from Berry and Houston (1995): (1) safe-driving speeds on newly resurfaced roads, (2) populations of rats in a confined space, or (3) the handicapping of weight lifters. The fourth option was a model created by a group for project 1 on the heart rate of mammals. Three groups worked with the heart-rate model, whereas one chose the rat-population model and one chose the driving-speed model.

In all groups, the quality of the writing on project 2 improved from that on project 1. Every paper contained a discussion of the problem. The students included mathematical calculations, assumptions, and facts in their paragraphs. However, the fact that these written projects included less mathematical analysis than did the previous projects may have accounted for the improvement in the quality of the writing.

Projects 3 and 4 required students to create a model to solve a specific problem, validate any assumptions they made, and analyze the model. Each group selected a different problem. Among those chosen was the problem "How far can a migrating bird fly without food?" (Bender 1978).

That team that chose the migrating-bird problem started by researching bird migration. They determined that the fundamental problem was that the birds do not use energy at a constant rate. The group noted that "as the bird's weight decreases, the bird should become more efficient at flying, and the rate of fat being used should decrease... . The relationship between weight and time is not linear but some type of curve." By building on the familiar "distance=rate*time" formula, the team created the following model:

$$D=R^*(I-E)^*K^*(1/Z)^*(1/W(t)),$$

> where $D$ is distance in miles, $R$ is rate in miles per hour, $I$ is premigration weight in grams, $E$ is postmigration weight in grams, $K$ is 9 kilocalories per gram of fat, $Z$ is a constant representing energy used per gram of body weight per hour, and $W(t)$ is bird weight in grams as a function of time.

The group members were confident in all parts of the model except the function $W(t)$. They included a graph of what $W$ should look like and made statements about the first and second derivatives of $W$. However, they were not able to develop an algebraic formulation for the function, even though their description was clearly of a decaying exponential function; and as a result, they doubted their model. The students modified the model to look at discrete values of $W(t)$ and tested this model using data on migratory birds. They found that the value of $Z$ does indeed seem to vary with different-sized birds. They also found evidence to support their idea that birds fly with constant effort, making $R$ a constant as well. The students acknowledged the magnitude of changing to discrete $W(t)$ values, saying, "this would yield an overestimation of the bird weight.... In our opinion, it is better to underestimate [flying distance] than overestimate it."

In all the students' papers for projects 3 and 4, their assumptions described the construction of the model. Thus, the mathematics was appropriate to the problem and followed from the assumptions. In all papers, the students clearly described a solution that related to the original question. They weighed their answers against common sense and actual data. The students used graphing utilities to generate curves to model their data. They also compared extrapolated values from models against measured values to determine an appropriate domain for the models.

As they progressed through all the projects, the students gained confidence in addressing subproblems. The students' discussions of assumptions and choice of mathematics tools improved during the semester. Although we can attribute some of this improvement to practice in writing reports on their models and incorporating the instructor's comments, we also think that the improvement can be attributed to the students' focus on specific aspects of the modeling process. As the course progressed, the instructors told the students precisely which areas of their process needed improvement.

One sticking point throughout the projects, however, was that most groups were unable to reflect objectively on the model they had created. For example, on one project, the groups stated the following:

- "These are valid items to consider when constructing a model of an epidemic but can add many complications to the task of finding a solution."
- "The assumption about all forms of milk having the same viscosity and density as skim milk may not be good because babies drink other forms of milk that are very different from skim milk.... If we were to do this again, we would test with more than one form of milk."
- "With more data, one would be able to write a model describing the relationship between weight and time."

In none of these examples do the students give a specific statement about how including more parameters or more data would improve the model. Rather, the students possessed only a vague notion that more is better. Including more class discussions on making assumptions and reflecting on the models might have helped the students identify specific changes that could be made. Overall, however, throughout the semester, the students' ability to create and understand mathematical models improved.

# CONCLUSION

BECAUSE THE course described above was an upper-level undergraduate mathematics class, it was not possible for us to follow the students into their teaching careers to assess outcomes in their own classrooms. However, our course evaluations showed that 67 percent of the students believed that the course content was "excellent." Anecdotal evidence supported this idea, and several students commented that they had acquired a greater appreciation for the complexity involved in teaching modeling and using open-ended problems. Such experiences appear to help prospective teachers develop skills in setting up rich mathematical problem-solving situations that integrate mathematics with other disciplines. We believe and our colleagues concur that a course like the one described above is integral to a program of studies for students majoring in mathematics teaching. As a result, we have institutionalized the course as a requirement in our school.

## BIBLIOGRAPHY

Austin, Joe D., James Hirstein, and Sharon Walen. "Integrated Mathematics Interfaced with Science." *School Science and Mathematics* 97 (January 1997): 45–49.

Beal, J., Dan Dolan, Johnny W. Lott, and J. Smith. *Integrated Mathematics: Definitions, Issues and Implications.* Helena, Mont.: Montana Council of Teachers of Mathematics, 1990. (ERIC Document Reproduction no. ED 3477071)

Bender, Edward A. *An Introduction to Mathematical Modeling.* New York: John Wiley & Sons, 1978.

Berry, John, and Ken Houston. *Mathematical Modelling.* London: Edward Arnold, 1995.

Burghes, David N., and Ian Huntley. "Teaching Mathematical Modeling—Reflections and Advice." *International Journal of Mathematical Education in Science and Technology* 13 (December 1982): 735–54.

Burke, Maurice, and Johnny W. Lott. "SIMMS Curriculum Development Philosophy." In *Monograph 1: Philosophies.* Missoula, Mont.: The SIMMS Project, 1993.

Camerlengo, Vivian M. "Mathematics Specialist–Teacher Program: An Intervention Strategy for All." In *Reaching All Students with Mathematics,* edited by Gilbert Cuevas and Mark Driscoll, pp. 119–131. Reston, Va.: National Council of Teachers of Mathematics, 1993.

Castle, Kathryn, and Douglas B. Aichele. "Professional Development and Teacher Autonomy." In *Professional Development for Teachers of Mathematics,* 1994 Yearbook of the National Council of Teachers of Mathematics (NCTM), edited by Douglas B. Aichele and Arthur F. Coxford, pp. 1–8. Reston, Va.: NCTM, 1994.

Cuevas, Gilbert, and Mark Driscoll, eds. *Reaching All Students with Mathematics.* Reston, Va.: National Council of Teachers of Mathematics, 1993.

de Lange, Jan. "The Teaching, Learning, and Testing of a Mathematics for the Life and Social Sciences." In *Applications and Modelling in Learning and Teaching Mathematics,* edited by Werner Blum, J. S. Berry, Robert F. Biehler, I. D. Huntley, G. Kaiser-Messmer, and L. Profke. Chichester, England: Horwood, 1989; New York: Prentice Hall, 1989.

Galbraith, P., and N. Clatworthy. "Beyond Standard Models—Meeting the Challenge of Modeling." *Educational Studies in Mathematics* 21 (April 1990): 137–63.

Holmes Group. Tomo*rrow's Teachers: A Report of the Holmes Group.* East Lansing, Mich.: The Homes Group, 1986.

Lester, Frank K. "Methodological Considerations in Research on Mathematics Problem-Solving Instruction." In *Teaching and Learning Mathematical Problem Solving: Multiple Research Perspectives,* edited by Edward A. Silver, pp. 41–69. Hillsdale, N.J.: Lawrence Erlbaum Associates, 1985.

———. "Musings about Mathematical Problem-Solving Research: 1970–1994." *Journal for Research in Mathematics Education* 25 (December 1994): 660–75.

MCTM/SIMMS. *Integrated mathematics: A Modeling Approach Using Technology.* Level 1–6, vol. 1–3. Needham Heights, Mass.: Simon & Schuster Custom Publishing Co., 1996–98.

Oke, K., and A. Bajpai. "Teaching the Formulation Stage of Mathematical Modelling to Students in the Mathematical and Physical Sciences." *International Journal of Mathematical Education in Science and Technology* 13 (December 1982): 797–814.

School Mathematics Study Group. *SMSG Unit Number Two,* Chapter 3 and Chapter 4. Stanford, Calif.: Stanford University, 1968. (ERIC Document Reproduction no. ED 222 341)

Swetz, Frank L., and J. S. Hartzler, eds. *Mathematical Modeling in the Secondary School Curriculum: A Resource Guide of Classroom Activities.* Reston, Va.: National Council of Teachers of Mathematics, 1991.

Thompson, Alba G. "Teachers' Conceptions of Mathematics and the Teaching of Problem Solving." In *Teaching and Learning Mathematical Problem Solving: Multiple Research Perspectives,* edited by Edward A. Silver, pp. 281–94. Hillsdale, N.J.: Lawrence Erlbaum Associates, 1985.

# Part 3

## Classroom Examples of Integrated Mathematics

11

# "The Octopus's Garden": A Case Study

Kathleen Hotaling-Bollinger

ONE OF my earliest attempts to integrate all areas of the curriculum at the early elementary school level occurred in May and June 1978 with a unit based on the song "The Octopus's Garden." The song was part of the *Silver Burdett Music* series (Aubin et al. 1974) that I was using to teach first grade, and an underwater theme seemed like a good escape from the summer's heat. I made waves and bubbles on the windows with blue and green tissue paper, and the class made a five-foot-high papier-mâché octopus to hang on our biggest bulletin board. Around the octopus, I hung construction-paper fish and other sea creatures that held instructions for related independent and small-group activities.

The unit spawned a series of activities that interwove all the subjects in the curriculum. The students learned the song and used it in music activities. They wrote additional verses and recorded them on cassette. We read fictional and nonfictional books about sea life and sailing, and the students wrote their own short stories and variations. I had reached map skills in our social studies curriculum, so I asked the students to make picture maps and maps using symbols to represent the octopus's garden. The students also used maps showing bodies of water to learn to use a compass rose and to identify real places on the earth. In science, the students learned about the ocean as an ecosystem, identified various sea plants and animals, and learned about properties of water and the water cycle. I integrated mathematics by using teacher-created story problems related to underwater life and by using fish-shaped crackers and other manipulatives. Because the students were working on place value and regrouping, I started with eight as our "magic number." After a week, I introduced the children to my "Decopus," which had ten tentacles and, as I told them, wanted groups of ten connected. The students used the Decopus as a springboard for learning about tens and hundreds. Art projects included fish mobiles made from plastic rings used to hold together soda six-packs, crayon-resist paintings (in which the artist first draws and colors the scene heavily in crayon and then paints over the scene with watercolor), and corallike sculptures made from synthetic packing peanuts. I took the class on a field trip to the New York Aquarium to observe sea life first-hand. I even had the students apply our underwater theme to movement in the room—students were allowed only to "swim" from place to place or to move as if they were wearing scuba equipment.

## EFFECTS ON TEACHING

PLANNING TO teach such an integrated unit made the use of the standard plan book frustrating because the topics and lessons overlapped. Because I was new to integrating subjects and such integration was not a trendy thing to do back in 1978, I tried to use those planning book boxes but ended up writing the same activities in several of them. As my principal became more used to what I was doing, she accepted a lesson plan for the unit as a whole followed by daily activity plans instead of requiring the more traditional subject-area breakdown.

Because the teachers had a curriculum to follow, even back in those pre-Standards days, I typically broke down the curriculum for the year into topics to be covered each month and noted page numbers and materials to be used with each topic. When each topic was finished, it was finished. Although I periodically referred to the material already covered, it had not occurred to me to consider any subject other than science as spiraling throughout the year. In particular, mathematics had always seemed clearly sequential to me. I would teach subtraction after addition. I might divert briefly to teach graphing, geometry, time, or money but then would return to teaching addition with regrouping, and so forth.

As I developed thematic units, however, the mathematics topics, or strands, turned out to be woven throughout all of what I taught. I began by writing story problems for addition and subtraction with regrouping and by using the "decopus." But mathematics skills came into play in many other ways. In the mathematics center, I distributed "go fish" cards, undersea board games, ocean-related puzzles, sea-life-shaped lace-your-answer cards, and the like. As the unit progressed, the students also used colors, shapes, comparative sizes, and patterns to discuss science and social studies concepts and observations. They graphed types of plants and animals seen at the aquarium as well as on posters and in illustrations. The students used charts to keep track of who had completed various mathematics-center-based activities. They measured distances on maps and measured volume and weight when learning about properties of water. The students dealt with probability and uncertainty when they predicted what they might see at the aquarium and then afterward checked their predictions chart. Because we had a schoolwide mathematics program and the students were expected to complete most of the textbook, they transferred the skills and concepts developed while working on the "Octopus's Garden" unit to the situations described in the textbooks. My idea of mathematics as a subject area made up of discrete topics had changed significantly.

The students successfully learned the skills and concepts taught in the unit. To evaluate their progress, I used teacher-developed tests in conjunction with the typical mathematics test of the time. The latter test featured the familiar combination of ten computation problems and five word problems. Everyone passed, with most students scoring above 85 percent. (Whatever the current passing grade might be, I do not consider that I have done my job if the children do not achieve at this level. I would be seriously reluctant to take myself to doctors, my car to mechanics, or my clothing to dry cleaners who knew just 65 percent of their field's techniques and information.) At that time, my school system was using the Metropolitan Achievement Test, and my students scored well on the end-of-year test. Many second-grade teachers spoke positively about the incoming students' preparedness, and none complained.

## INTO THE CURRENT MILLENNIUM

THE EXPERIENCE of teaching the "Octopus's Garden" unit more than twenty years ago changed my thinking—and my teaching. Now, with my integrated teaching experience as well as new technology innovations, I find that integration is an even more powerful tool. Even with new standards and requirements, integration can be used effectively to bring about in-depth learning and true understanding.

One difficulty in thematic teaching has always been the need to use a given school's mathematics program. This constraint is especially true in today's urban schools, where districts trying to improve standardized test scores require their teachers to use very specific techniques and materials. Unfortunately, the provided textbook and workbook materials rarely fit a theme. The school in which I teach is not "under registration review" (which in New York means district-level, rigorous supervision geared toward improving students' performance), and it is part of a much smaller system than are most urban public schools. Therefore, as long as I cover the New York state standards for the grade level that I teach, I have the freedom to create my own activities and worksheets and to incorporate the school's mathematics program as independent practice or homework.

However, in schools where such flexibility is not possible, good mathematics can still be woven into themes based on science or social studies topics, just as it was woven "under the sea" in the "Octopus's Garden" unit. Such integration can be done while following to the letter, and as scheduled, the school's reading or English Language Arts program and its mathematics program. The concepts that the students learn can be nicely applied in the time allotted to science and social studies because they are the easiest areas in which to incorporate a thematic approach.

Support materials other than the textbooks are a source of delight. Although we used fish-shaped crackers in my unit, teachers can readily use plastic manipulatives. (Note that craft stores are a much cheaper source of supplies than are teachers' supply stores.) And even though objects that look like part of a theme are fun to use, such standard items as connecting cubes and Cuisenaire rods can also be easily used. Workmats—pieces of paper that define each student's space for working with manipulatives—are easily and inexpensively reproduced or created and decorated by the students and can be combined with the aforementioned tools to good effect. This hands-on approach also applies to teachers of upper grade levels, who should not limit themselves to books and calculators. Mirrors, lines of symmetry, and hieroglyphics can be used effectively, as can historical weather data compiled in charts and graphs that students can use to make predictions in a unit on climate and its impact. Students can also use such items as pieces of straws to explore relationships in right angles and apply those relationships to architectural construction in a unit on architecture, ancient civilizations, and urban living.

In the last ten years, the technology explosion has been incredible—and has offered teachers a number of valuable tools for teaching and integrating ideas. Back in 1978, no one would have allowed first-grade students to use calculators. Today, calculators are a way of life. I have younger students use calculators for skip counting, and those students can demonstrate the appropriate use of calculators in problem solving. Today, I would use calculators in the "Octopus's Garden" unit for having the students solve word problems and calculate the number of creatures involved in short narratives or seen in illustrations and on posters. In calculation-skill activities, calculators can nicely be used with student-pairs who are trying to "beat the calculator" in performing computations.

The use of the computer in the classroom has had an even more dramatic impact. I am amazed to think back to a time when classroom computers did not exist. To bring my twenty-year-old "Octopus's Garden" unit to the new millennium, I would have the students use the Internet and instructional software to research marine life. Children can also use the KidPix software program to illustrate number concepts and to compose their own story problems, write the number sentence to solve them, and then draw those problems.(See figs. 11.1 and 11.2.) Students can scan photographs and add text to make number books. These activities are staples in my class today. I also use the Graph Club software program to keep a record of the students' graphs. That record becomes a book at the end of the unit. The students have the option of using word-processing software for their stories, poetry, and reports. If I were doing the "Octopus's Garden" unit, one culminating activity would be creating a multimedia presentation showing what the students have learned about life under the sea.The presentation would include information about the plants and animals that live in the water here in our local area. I would then have the students share that information with a class of e-mail pals from a dissimilar region. Such exchanges could be posted on the school's Web site.

Fig. 11.1 In completing the assignment of writing their own story problem using a popular software program, the students practiced operations, explored number and numeration concepts, engaged in mathematical reasoning, and employed modeling and multiple representations.

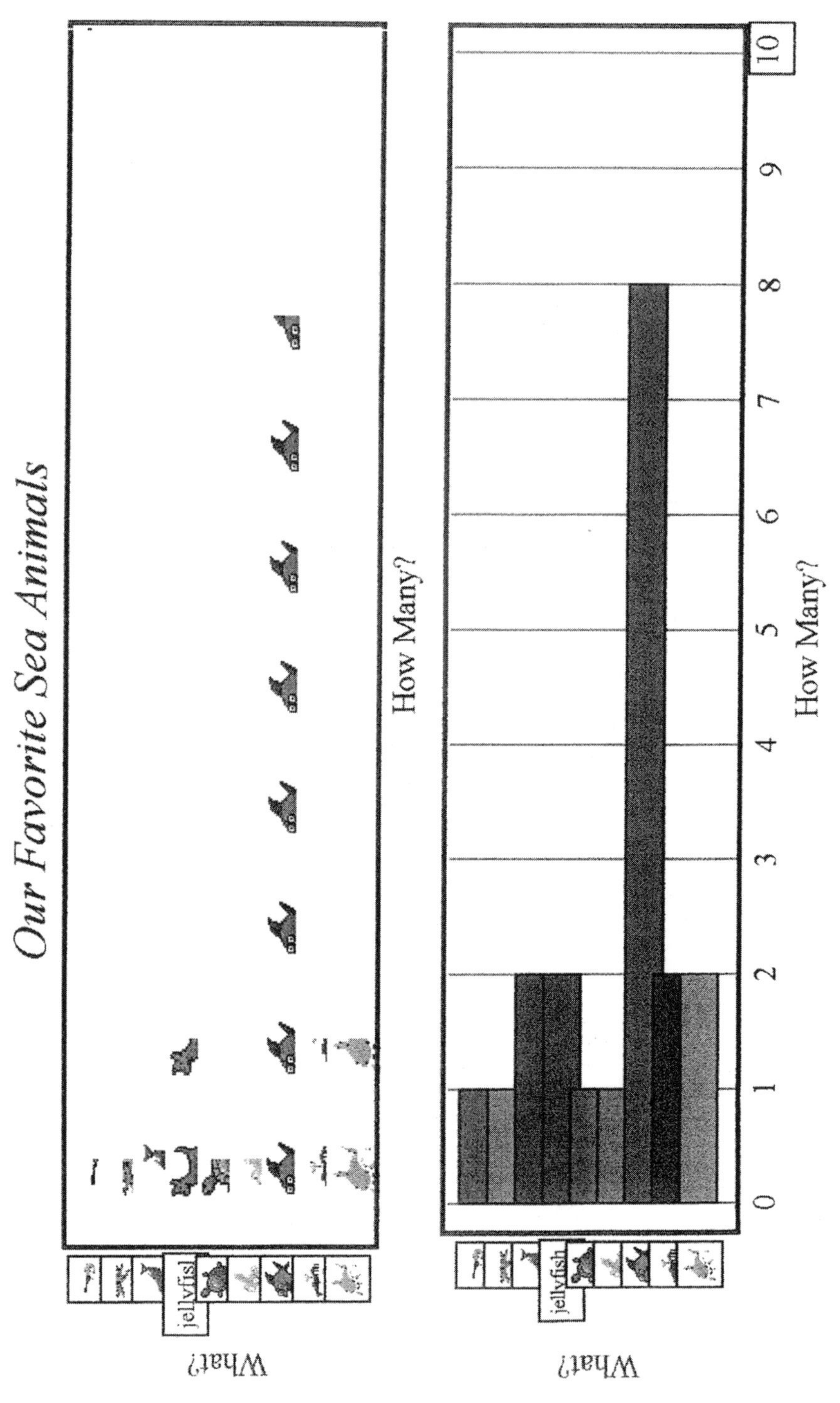


Fig. 11.2. Our class graph of favorite sea animals, created using a commercially available graphing program; the electronic record of students' graphs became a book at the end of the unit.

## STANDARDS-BASED PLANNING AND STANDARDIZED TESTING

STATE STANDARDS and a strong emphasis on standardized testing are the realities of the times in which we teach. In no way do they conflict with an integrated curriculum. The requirement, however, is thorough planning.

To navigate the standards and testing requirements, I keep an electronic spreadsheet that contains the standards for each curriculum area for my grade level. As I cover each standard in a unit, I note the activity in the appropriate box. This record keeping enables me to cover all topics and show my coverage of the standards to the principal, curriculum coordinator, district personnel, parents, and anyone else who needs to know. My spreadsheet also makes it easy to create classroom signs indicating the standards being worked toward at any given time. I like to use Inspiration software to plan my units because it enables me to see the full spectrum of the curriculum on one page. Typically, I put science and social studies standards on the diagram page, and I note mathematics and language arts standards in my outline because these concepts and skills are repeated across the curriculum. For the purposes of this chapter, I have placed the mathematics standards and activities on the sample chart shown in figure 11.3. In addition, I create a skills, processes, and concepts chart (see table 11.1) for each unit. Finally, I use a matrix that ensures that I cover the levels of thinking in Bloom's Taxonomy (Bloom 1956) and that I address multiple intelligences (see table 11.2).

I use a number of activities to keep my students focused on their own progress. I require my students to tell me one thing that they have learned each day. As they share what they have learned, I help them understand which subject area that they are talking about. The students also have "learning journals," in which they finish such sentences as "Today I used mathematics when I ____." These entries vary from day to day but help the children see the ways that the various subjects and learning styles interact.

When it is time for assessment, I use both formal and informal methods. For my purposes, I find that the informal methods are more useful. However, the students—even if they are too young for the state tests, as mine are—need to be comfortable with the format of the standardized test when the time comes.

Informally, I ask the students to tell me about their projects. These projects include writing, illustrations, artwork, models, journals, self-

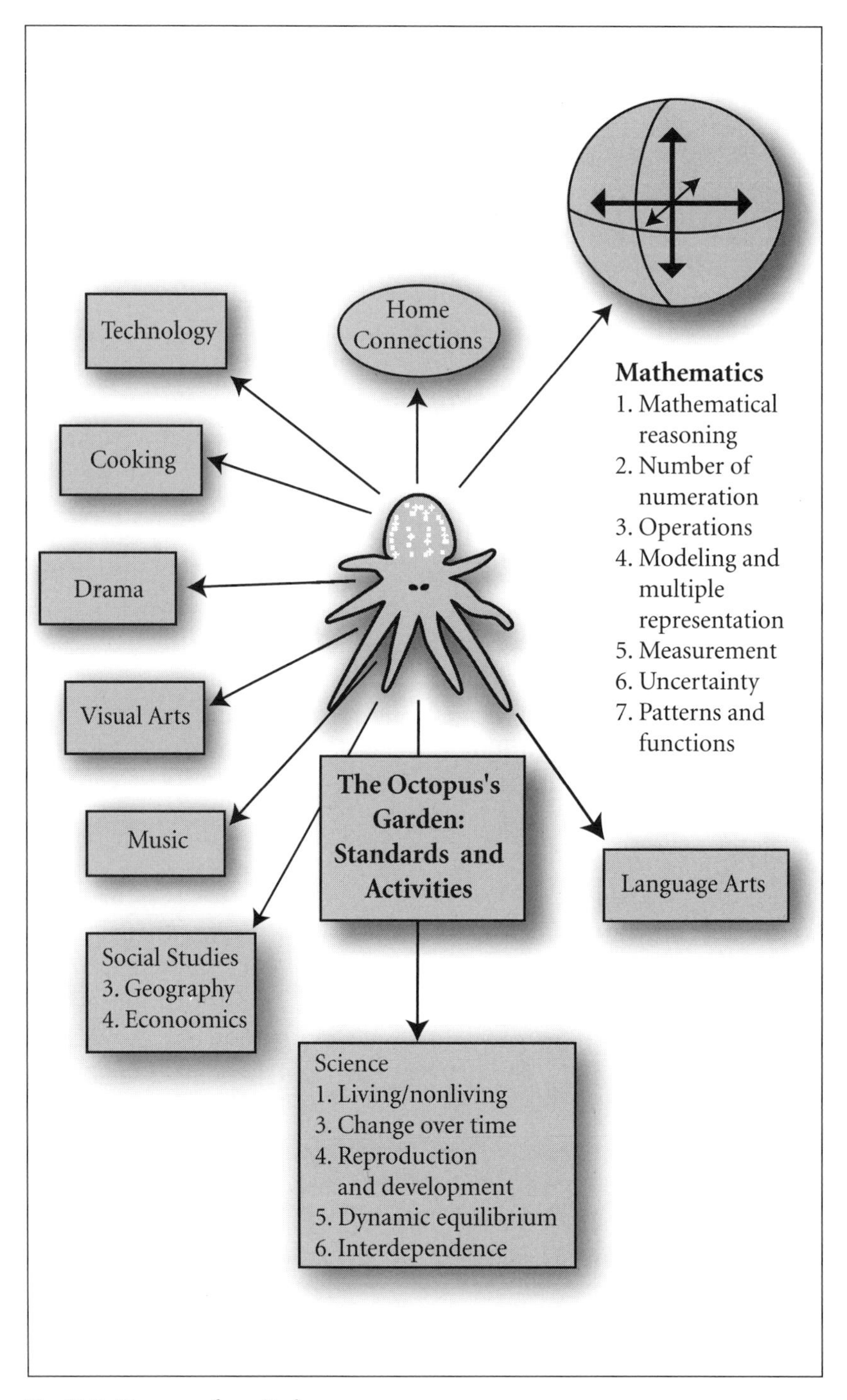


Fig. 11.3. Diagram of a unit plan

evaluations of cooperative-learning activities, and computer work, either printed or on disk. The questions I might ask include the following:

- What animals are here?
- What do they do in the ocean?
- How do they eat or protect themselves?
- Which animals are the biggest or smallest? How do you know?
- What can you tell me when you interpret this graph?
- Is there a pattern here? Explain it to me.
- How did you know which number sentence to write?
- When you wrote about this, what letters did you hear?
- Where is an ocean on this map?
- How do you know?
- Why can't this animal live on land?
- Do we eat any of these? Do you like them?
- How did you find that out?"

I also ask the children to tell me how the project was done, what was the best part, what was the hardest part, or what part they didn't like, and why. I ask them what they consider the most important thing that they learned and to tell why. I ask if they still wonder about anything with regard to the project. I make notes about their writing, reading, and expressive language skills. I ask the students to demonstrate such mathematics skills as measuring an object, showing a specific number, using a visual model, interpreting a graph, telling how to solve a story problem, solving a story problem, and so forth.

A formal test would include computations as well as problem solving with explanations. This year, I am teaching kindergarten. Therefore, I would include five computation-type questions (e.g., "Draw a ring around eleven fish" or "5 + 2 =___") and one problem to solve ("Show me which one of these things happened last, and tell me how you know" or "If six seahorses were in the Octopus's cave and two swam out, how many would be left inside? Show how you got your answer."). Because they are kindergartners, I test the students individually on the problem portion so that their dictated answers can be written down. I also allow the students to draw their answers when appropriate. For older students, formal tests should be longer and the students would be expected to write how they arrived at their answers.

TABLE 11.1. Skills, Processes, and Concepts Chart for "The Octopus's Garden" Unit

| State Subject-Area Standards | | | | |
|---|---|---|---|---|
| **Science** | **Language Arts** | **Mathematics** | **Activities** | **Assessment** |
| (1) Living/nonliving things | (1) Listen/read/speak/write for information (acquire and classify information, manipulate ideas, record data) | (1) Mathematical reasoning (4) Modeling and multiple representation (5) Measurement (7) Pattern and function | (a) Sort pictures and models of things to be found in the sea into alive/used to be alive/never alive (b) Sort water animals into ocean-dwelling/other (c) Identify animals in calendar pattern (d) Use fish-shaped crackers as a nonstandard unit of measure | Activity should show ability to identify characteristics and sort by them. |
| (3) Organisms change over time (4) Continuity of life is sustained through reproduction and development | (1) Listen/read/speak/write for information (acquire and classify information, manipulate ideas, record data) (2) Listen/read/speak/write for literary response and expression (present reactions, write poetry, choose favorites) | (5) Measurement (comparing sizes at different stages of life, comparing sizes of different animals, comparing lengths of life spans) | (a) Use software and books to learn about life cycle of chosen animal (b) Learn songs about growth and change among sea animals (c) Read stories and poetry about growing animals (d) Prepare a multimedia presentation showing the life cycle of chosen sea animal (e) Make a class big book of sea animals (f) Write poetry: "The ____ and I" | Presentation shows whether animal is oviparous or live-born, what it looks like, and what it is called as a baby and an adult. |

TABLE 11.1. Skills, Processes, and Concepts Chart—Continued

| State Subject-Area Standards | | | | |
|---|---|---|---|---|
| **Science** | **Language Arts** | **Mathematics** | **Activities** | **Assessment** |
| (5) Organisms maintain a dynamic equilibrium that sustains life | (1) Listen/read/speak/write for information (acquire and classify information, manipulate ideas, record data)<br>(2) Listen/read/speak/write for literary response and expression (present reactions, write poetry, choose favorites) | (1) Mathematical reasoning<br>(4) Modeling and multiple representation<br>(5) Measurement<br>(7) Pattern and function | (a) Use books, software, and videotapes to learn about habitats and foods<br>(b) Build a diorama or model ocean habitat<br>(c) Paint ocean pictures<br>(d) Sample some foods that ocean creatures eat<br>(e) Use ordinal numbers and position words to describe position in pictures and models<br>(f) Dramatize the book *The Rainbow Fish* and the song "The Octopus's Garden"<br>(g) Estimate the volume of various containers<br>(h) Estimate how many students or fish will fit in various "sea beds" | Models and pictures should visually or verbally show appropriate feeding patterns, shelter, and means of protection from danger; participation in dramatic presentations should show comprehension; appropriate mathematical vocabulary should be used. |

*(Continued on next page)*

TABLE 11.1. Skills, Processes, and Concepts Chart—Continued

| State Subject-Area Standards | | | | |
|---|---|---|---|---|
| **Science** | **Language Arts** | **Mathematics** | **Activities** | **Assessment** |
| (6) Interdependence | (1) Listen/read/speak/write for information (acquire and classify information, manipulate ideas, record data) (2) Listen/read/speak/ write for literary response and expression (present reactions, write poetry, choose favorites) | (1) Mathematical reasoning (2) Numbers and numeration (3) Operations (4) Modeling and multiple representation | (a) Dramatize *The Rainbow Fish* (b) Describe how ocean plants and animals are important to one another (c) Choose favorite animal at the end of the unit, and state why; graph the choices, and interpret the graphed data (d) Use story problems about ocean creatures and fish-shaped crackers to solve addition and subtraction problems (e) Use KidPix software to create own story problems using sea creatures (f) Use the "decopus" and fish-shaped crackers to demonstrate place value (g) Take trip to aquarium and write thank-you note | Correct operations are selected, and story problems are solved; original story problems make sense; understanding of place value is demonstrated accurately; oral descriptions contain at least three facts; and reference is made to models, dioramas, and multimedia presentations. |

TABLE 11.1. Skills, Processes, and Concepts Chart—Continued

| State Subject-Area Standards | | | | |
|---|---|---|---|---|
| **Social Studies** | **Language Arts** | **Mathematics** | **Activities** | **Assessment** |
| (2) Geography | (1) Listen/read/ speak/ write for information (acquire and classify information, manipulate ideas, record data) | (5) Measurement (6) Uncertainty | (a) Locate various bodies of water on the world map and discuss whether they are near or far (b) Compare sizes of various bodies of water on a map (c) Locate the aquarium on a map and identify nearest body of water | The student correctly identifies oceans, rivers, and lakes; appropriately uses "bigger," "smaller," "closer," "further;" explains reasoning. |
| (3) Economics | (1) Listen/read/ speak/write for information (acquire and classify information, manipulate ideas, record data) (2) Listen/read/ speak/write for literary response and expression (present reactions, write poetry, choose favorites) | (1) Mathematical reasoning (4) Modeling and multiple representation (5) Measurement | (a) Find out what sea creatures we use for food and decide which ones we want to eat (b) Create a seafood restaurant in the dramatic play area (c) Choose a water vacation, describe or paint it, and tell why it was chosen | The student names at least three seafoods; tells or shows how those foods are harvested; chooses from among foods and vacations and explains choice; makes changes. |

TABLE 11.2. Levels of Thinking and Multiple Intelligences Grid for "The Octopus's Garden" Unit

| | **Recall/ Compre-hension** | **Application** | **Analysis** | **Synthesis** | **Evaluation** |
|---|---|---|---|---|---|
| Linguistic/ Verbal | Recite poetry; listen to books | Show and tell about the habitat | Sort and classify | Compare/ contrast living and nonliving | Choose favorite and tell why |
| Mathemati-cal/Logical | Make graphs | Prepare graphic organizer (word web) showing vocabulary; solve story problems | Interpret graph data; compare sizes | Invent story problems; estimate size in yarn or crackers | Survey people's favorite sea creatures |
| Visual | Identify calendar pictures | Paint sea life | Illustrate animals in different habitats | Create paintings of your own octopus's garden | Choose favorite masterwork picture of the sea |
| Bodily/ Kinesthetic | Dramatize stories | Role-play restaurant scene | Categorize bodies of water by moving to them on large map | Build habitat | Choose a vacation |
| Musical | Sing songs about sea life | | Dramatize "The Octopus's Garden" | Write a song (or verses) about sea animals | Select recorded music to accompany presenta-tions |
| Intra-personal | Write journal | Help with KWL-H chart (what we Know, what we Want to find out, what we have Learned—How we learned it) | Describe activities at the beach | Imagine how it would be living under the sea | Describe the most impor-tant thing learned and why |

TABLE 11.2. Levels of Thinking and Multiple Intelligences Grid—*Continued*

| | **Recall/ Compre-hension** | **Application** | **Analysis** | **Synthesis** | **Evaluation** |
|---|---|---|---|---|---|
| Inter-personal | Listen to each other's presenta-tions | Work in teams on place-value activities | | Figure out what will be served at the restaurant | Describe the best and worst things that hap-pened on the aquari-um trip |
| Techno-logical | Find out facts about sea life | | Design sea scenes with drawing software | Create story problems with art and multimedia software; design a multimedia presentation | |

Adapted from a matrix provided by the Diocese of Brooklyn (1998)

# CONCLUSION

CHILDREN LEARN best when the learning is meaningful. With careful planning and thorough assessment, integrated teaching incorporates mathematics quite naturally. Integrating mathematics into the rest of the curriculum gives the mathematics a context. Furthermore, thoughtful integration helps to remove mathematics from the realm of tedious practice and place it firmly in the realm of essential and dynamic tools.

## BIBLIOGRAPHY

Aubin, Neva, A. Beer, J. Beethoven, Elizabeth Crook, and David S. Walker. *Silver Burdett Music.* Morristown, N.J.: General Learning Corporation, 1974.

Baratta-Lorton, Mary. *Mathematics Their Way.* Menlo Park, Calif.: Innovative Learning Publications, 1995.

Bloom, Benjamin S., ed. *Taxonomy of Educational Objectives: Handbook I: Cognitive Domain.* New York: Longman, 1956.

Diocese of Brooklyn. "Science Curriculum K–8: New York State Learning Standards: Science." Brooklyn, New York: Diocese of Brooklyn, 1998. Photocopy.

Diocese of Brooklyn. "Social Studies Curriculum Guide Kindergarten–Grade 4." Brooklyn, New York: Diocese of Brooklyn, 1998. Photocopy.

New York State Education Department. "English Language Arts Core Curriculum." Albany, NY: The University of the State of New York, 2001. http://www.emsc.nysed.gov/ciai/pub/pubmst.html (13 January 2003).

New York State Education Department. "Mathematics Core Curriculum." Albany, N.Y.: The University of the State of New York, 2001. http://www.emsc.nysed.gov/ciai/pub/pubmst.html (13 January 2003).

12

# Weaving the Threads of Mathematics Learning throughout Elementary-Level Education: A Look at Reconstructing Curricula

Mia Nazzaro Ramirez

THE POWER of an integrated, thematic curriculum is undeniable: Children make significant connections within and between subject areas because their focus is not fragmented, they are better able to solve real problems (Hankes 1996), and they recognize the value of their learning. These inherent benefits alone should be reason enough for us to dedicate ourselves to integrated learning. But, also consider the increasingly demanding national and state standards in each subject area. Particularly in the primary years in which building literacy is a focus, the time devoted to mathematics instruction is limited and expectations are high. Elementary school teachers struggle to give each subject fair representation in the fabric of a patchwork school day full of special classes, assemblies, and fire drills. For teachers, an integrated curriculum is not only beneficial for students' learning but also an essential, efficient way of maximizing precious educational time with students.

Once educators are committed to integrated learning, how do we put these ideas into practice? This chapter explores that practical question and describes an ongoing process of developing an integrated curriculum. The chapter focuses on how first-grade teachers at Glenwood Landing Elementary School in Long Island reconstructed the academic year around three thematic units: the ocean, folk and fairy tales, and woodland studies. It examines how the units addressed standards in all areas and introduced and reinforced specific first-grade skills. The chapter also includes guidance on developing a comprehensive folk-and-fairy-tale unit and on implementing a mathematics-oriented "Three Little Pigs" ministudy.

## GETTING STARTED: SIX STEPS TO INTEGRATION

SEVERAL YEARS ago, a diverse team of Long Island educators formed to consider the process of creating an integrated curriculum for early childhood education. The group, hailing from different areas of expertise, included Barbara Breslow, an early childhood center director; Wynne Shilling, codirector of the Literacy Enhancement Project in New York City; Elisa Barilla, a music teacher; and me, a first-grade teacher and mathematics specialist. As we shared our collective experience, we compiled the following steps for the effective development of an integrated curriculum.

### Step 1: Know the mathematics

As you begin to reconstruct your curriculum, the first step is to carefully review and gain a deep understanding of the Standards described in NCTM's *Principles and Standards for School Mathematics* (NCTM 2000), the state standards that apply to your school district, and any programmatic or school expectations. What are the specific mathematics objectives? What must students know and be able to do at the end of the year? Does your mathematics program spiral to revisit and build on several mathematical topics at once, or is it unit oriented? If you are new to your school district, how much flexibility do you have in deciding how and when your curriculum will be delivered? The answers to these questions are crucial because they represent the foundation of your program and serve as the criteria by which you and your students will be evaluated.

### Step 2: View the world through the lens of the mathematician

At the heart of integrated-mathematics learning is children's experience of mathematics as a vital, living science that opens minds to new ways of seeing and analyzing the world. For this experience to occur, our thinking as teachers must fundamentally change so that we recognize mathematics in our own world. As we remove the blurred lens of the curriculum generalist and put on the focused lens of the mathematician, we will see the incredibly rich opportunities for worthwhile investigations and activities that were there all along. Start by examining existing lesson plans in all subject areas. Where is the mathematics? What thematic units can bring out the mathematics to which children should be exposed at your grade level? Do you see a common thread that can be woven into a continuous web of thought? Allow ideas to arise spontaneously from the old to create the new, and use the answers to the preceding questions to decide on a theme or comprehensive unit.

### Step 3: Spread the word

Talk with colleagues about your exploration of new ideas and interest in an integrated curriculum. You might be surprised at the support available when you make your needs and interests known to those around you. Important people to include in your efforts are your mathematics specialist or curriculum associate as well as your school's librarian, art and music specialists, and physical education teachers. Expanding your network will lessen your burden and keep you motivated. Also try to find a colleague with whom you can collaborate and brainstorm possibilities—a kindred spirit can be immensely helpful in sustaining your efforts and in ensuring that you develop the best possible program for your students.

### Step 4: Research

Here is where you do your own homework on the topics that you want to cover. Use the resources available to you, such as the professional journals and educational materials in your school. Visit your public library to get a sense of the materials that will be available to your students for additional study and exploration. The Internet is another valuable resource through which to access diverse ideas. In online searches, often you can find professionals with similar interests and pursuits. Establish a file for research on your topic so that you can stay organized as your networking and research bear fruit.

### Step 5: Brainstorm

Now that you have a wealth of related activities and background information on your topic, use a curriculum web or graphic organizer listing all the subject areas to brainstorm the possibilities for a unit. Get an overview of the unit so far by considering which areas, if any, can or should be developed further. How can you enhance your unit considering all that you know is necessary for your grade level? Do you need to revisit teaching specialists? If so, give them your completed curriculum web and ask for suggestions. Perhaps they can adapt their own plans to supplement your curriculum.

"Where's the mathematics?" you should ask yourself. In this part of the brainstorming step, think about using thematic curricula to connect necessary concepts and to support previously discussed topics in mathematics. Also consider how to introduce and reinforce specific mathematical objectives. If you have done your job in "Step 1: Know the Mathematics," you can now thoughtfully seek out those integrated mathematical experiences that will help your students meet learning objectives. Next use a web sheet to brainstorm a more-specific list of possible activities or investigations. In which boxes do your ideas naturally fall? What, if any, other areas of mathematics need to be considered? When you look at possibilities wearing your mathematical lenses, what new opportunities jump out at you? Can you create a mathematical poem or activity that supplements a particular area lacking in your mathematics curriculum? If not, how will those needs be met? These mathematical curriculum webs become essential documents that will help you correlate the requirements of your program with your integrated curriculum.

### Step 6: Present worthwhile mathematical tasks

As mentioned previously, our time with students is limited. Next you must determine which of the brainstormed ideas are most worthwhile for your students. Use Standard 1: Worthwhile Mathematical Tasks in *Professional Standards for Teaching Mathematics* (*Professional Teaching Standards*) (NCTM 1991, p. 25) to further select and refine activities. In essence, students should be presented with activities that represent significant mathematics in meaningful, engaging contexts and that give students of all learning styles opportunities to learn.

As you consider how well a given mathematics program meets the NCTM's criteria, an additional question must be considered: Can you do more to align the program with the ideals in *Professional Teaching Standards*? One way to find out is to look at a familiar lesson through your new mathematical lenses. We did, and we found new ways to help our students make connections. For example, we looked at one lesson from our first-grade Comprehensive School Mathematics Program (CSMP) (McREL 1991); the lesson is summarized below.

> Introduce $2x$ in a story about doubling. Label the dots in a $2x$ arrow picture, and use a calculator to check some of the calculations.

The scripted lesson continues with a story about children's bringing a new friend to school each day. At the time, we were reading Chinese folktales and were about to read *Two of Everything* (Hong 1991). The idea that this folktale could be used to introduce and explore the concept of doubling immediately became apparent. Accordingly, during the second reading of the story, we captured the children's imagination by modeling the doubling function with a brass pot. Using the pot furnished an opportunity for concrete exploration of the concept that the original lesson had not presented. Many students in first grade needed that added experience to grasp the difficult concept of doubling. As an extension, the class made a book featuring the students' answers to the prompt "I put ______into the pot, and out came _______." The children included both addition and multiplication number sentences to accompany their drawings. In this example, viewing the old curriculum through new lenses helped us as teachers recognize a more worthwhile opportunity that met a specific mathematics objective.

As our folk-and-fairy-tale unit continued, the children repeatedly read *The Gingerbread Man* (Aylesworth 1998) to gain reading fluency and phrasing. We adapted one of our CSMP lessons for use with the story, as seen below.

> *Original lesson*:
> Determine how many days are in a week. Show how long a week is on a calendar. Discuss how many days are in a month. Tell a story that poses the problem of calculating $5 \times 7$. Draw a picture, and use the minicomputer to calculate $5 \times 7$. [Scripted story about a girl visiting her grandparents on the farm followed.]
>
> *Revised lesson*:
> The gingerbread man ran away for five weeks. How many days was the gingerbread man away altogether?

This revised question led to the same inevitable conversation about how many days are in a week, and the specific techniques for problem resolution mentioned in the CSMP program were explored. Unlike in the original lesson the year before, however, the classroom conversation surrounding the "gingerbread man" version was livelier and the children demonstrated a greater sense of commitment to the task because they were more invested in the context of the problem. Although this kind of activity does not take the place of more authentic problems that naturally arise from the study of a subject, it does help children to make connections among the otherwise fragmented segments of their day. At times when I have felt conflicted about this type of adaptation, I have thought, "What ends could have been achieved by strictly following the original version of the lesson?" The answer is virtually none.

# A MINISTUDY: THE POTENTIAL OF "THE THREE LITTLE PIGS"

WORKING WITH Debbie Minicozzi, a neighboring teacher and mathematics enthusiast, the first-grade teachers at Glenwood Landing Elementary School, where I teach, developed a ministudy based on "The Three Little Pigs." We believed that the ministudy would maximize our instructional

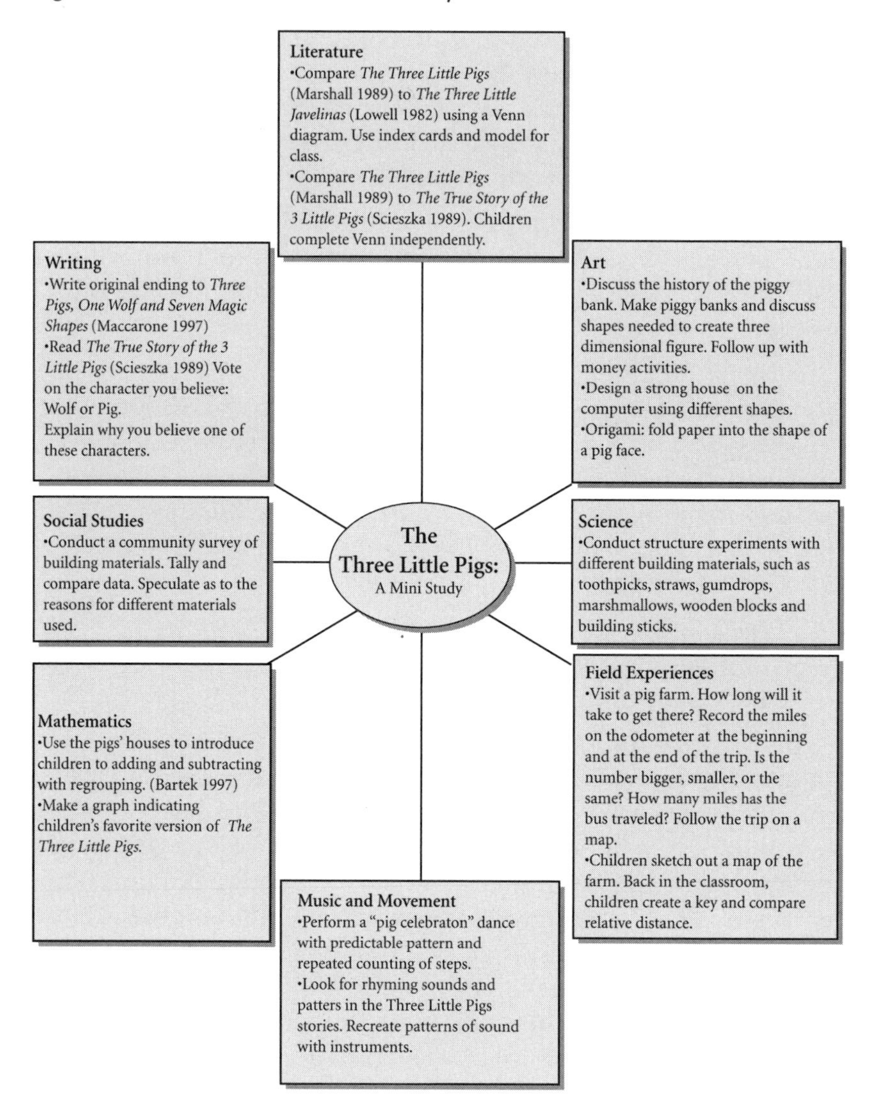


Fig. 12.1. A sample thematic study by subject area

time and give students contextualized learning in mathematics. Using the process outlined previously, we discovered an incredible wealth of opportunities and unexpected benefits. Beginning with a curriculum web, we brainstormed activities that related to different subject areas as well as contained mathematical elements (see fig. 12.1). We studied the NCTM's Standards for prekindergarten through grade 2 (NCTM 2000, pp. 72–141) and continued brainstorming related activities for each Content Standard (see fig. 12.2). We eventually selected the most worthwhile

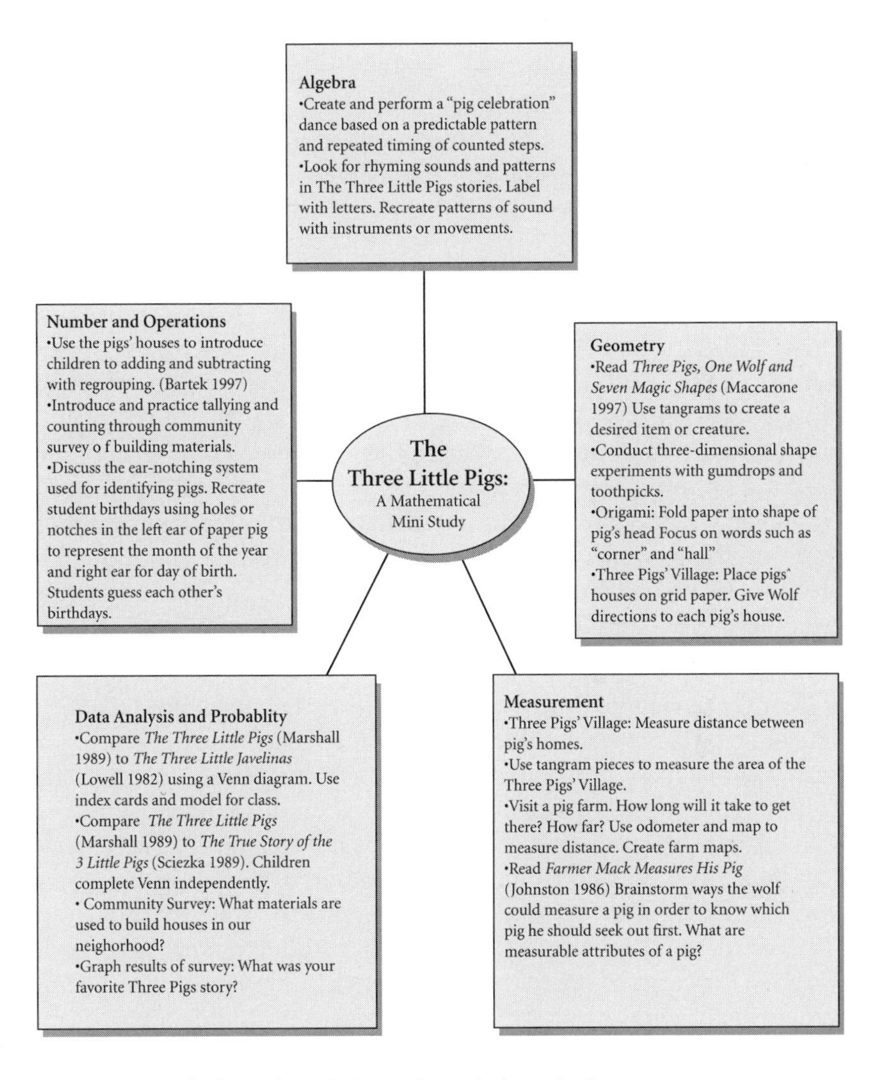


Fig. 12.2. A sample thematic study by mathematical standard

activities that would meet specific first-grade objectives. The following is a description of a few activities that we developed as well as exciting discoveries that we made about our students, the curriculum-development process, and ourselves.

## Venn Diagrams: A New Beginning

Perhaps one of the most obvious ways of integrating mathematics into a literature-based, fairy-tale unit is through the use of Venn diagrams. Although Venn diagrams offer a wealth of opportunities to analyze data, introducing them to young children can be a challenge. In first grade, we began by assigning the children different aspects of two versions of the story and giving them corresponding colored index cards: yellow if they were describing a story element found only in *The Three Little Javelinas* (Lowell 1992), blue if they were describing a more traditional version of *The Three Little Pigs* (Marshall 1989), or green if they were describing the similarities between the two versions. As the children came together to share their observations and to place them on the diagram, a pattern immediately began to form. They were animated when predicting and confirming the pattern as each card was correctly placed. The carefully constructed activity also allowed for more-advanced observations. A few children noticed that when the colors blue and yellow are combined, they create the color green—a further representation of the diagram's meaning. The diagram from this lesson served as a model for students and allowed them to begin completing Venn diagrams on their own as they reflected on the similarities and differences between different "Three Pigs" versions, such as *The Three Little Pigs* (Marshall 1989) and *The True Story of the Three Little Pigs* (Scieszka 1996). See figure 12.3 for an example.

In interweaving literature, data analysis, and pattern recognition, we were able to revisit essential mathematical concepts during our regular literacy block. As a result of integration, Debbie Minicozzi found that her first-grade students were more aware of mathematical connections in literature. Specifically, her students began to look for such mathematical elements as repeated use of number, opportunities for story analysis as described above, and mathematical labeling of patterns. These elements also began to surface in her students' writing about literature. Our colleagues agree that mathematical integration—in which students engage in substantial literary discussion while practicing embedded, meaningful mathematical skills—has been an effective way to spend our time.

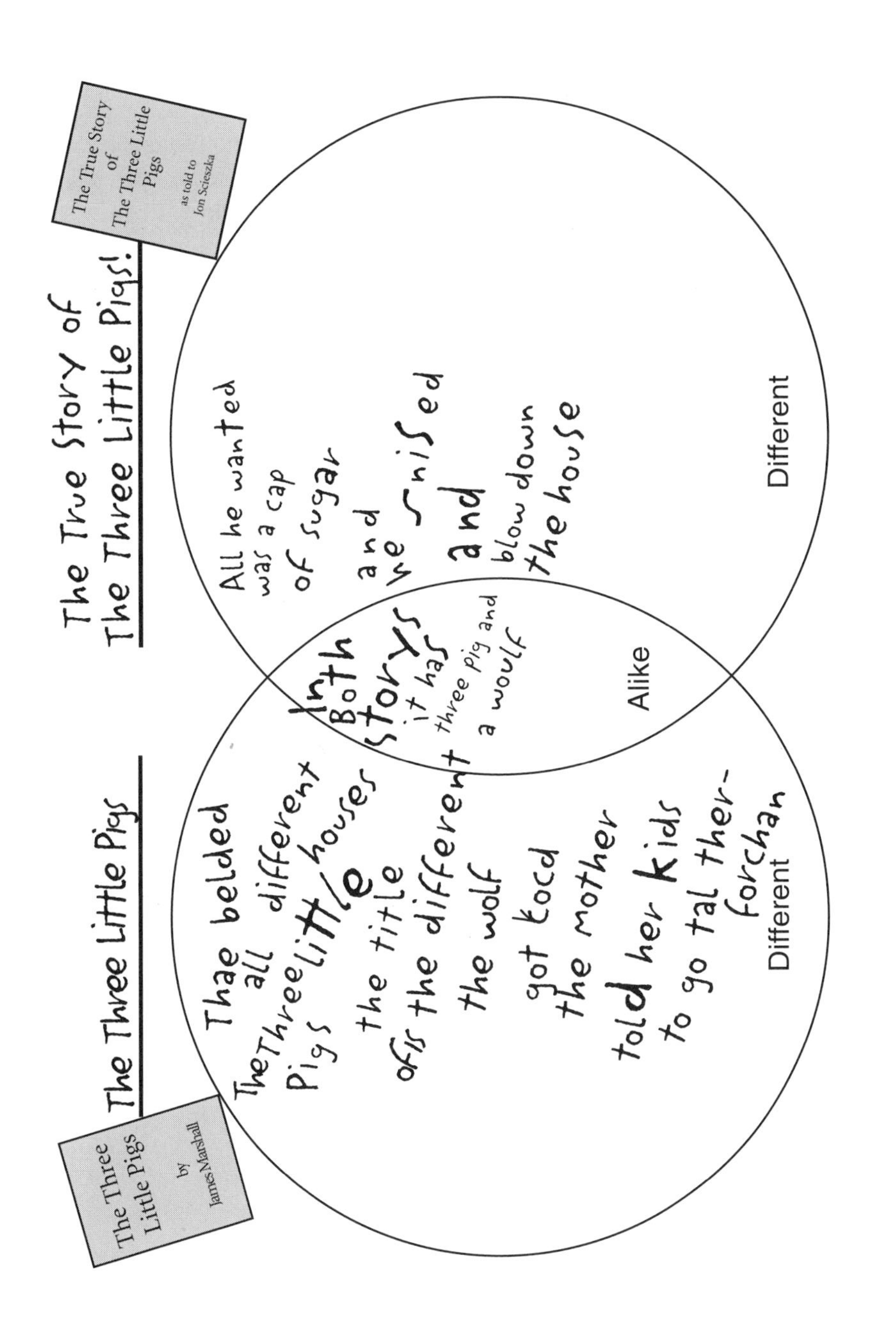


Fig. 12.3. Student sample of Venn diagram comparing "Three Pigs" stories

## Pigs and Polygons

As our "Three Pigs" unit developed, we were looking for ways to meaningfully integrate geometry into our thematic study. In reading the story *Three Pigs, One Wolf, and Seven Magic Shapes* (Maccarone 1997), I was immediately struck by the opportunity to engage students in using visualization and spatial reasoning. In this story, each pig goes out to seek its fortune and is given seven magic shapes, or tangrams, which it is advised to use wisely. After the first two pigs meet an untimely end, I encourage the students to imagine that they are the third pig. The students write a new ending by completing the idea "If I were the third little pig, I would...." I also give the students a set of paper tangrams that can be used to make an item to help them get away or be safe from the wolf. The resulting work over the past three years has been remarkable (see fig. 12.4).

Figure 12.4. Students create tangram designs in response to a story.

Although the "Pigs and Polygons" lesson effectively combines creative story writing, prediction, and geometric manipulation of shape, the lesson offers more than efficient use of time. Before they began to write, the students were given time to visualize their thoughts, an opportunity that is essential for most children (Bell 1991). Amazingly, as the students started working, those children who normally had difficulty initiating the writing process became intently involved in their creations. This integrated mathematics writing tool aided those students because it engaged them and kept them on task throughout the activity. In addition, it challenged the children to mentally hold onto the gestalt, or whole geometric image, of their desired form as they flipped, turned, and slid polygons into place. Furthermore, this activity presented a means of expression for children with different learning styles. Kinesthetically inclined students were particularly animated throughout the assignment. Other students wrote feverishly and gained geometric problem-solving practice as they struggled to give life to their vision using their paper pieces. In looking at this activity, I am reminded of the many benefits of varying activities to reach children of diverse interests, learning styles, and abilities.

As our curriculum development of the geometry strand continued, we realized that additional areas needed attention. In addressing the expectation in *Principles and Standards* that students in prekindergarten through grade 2 be able to "specify locations and describe spatial relationships using coordinate geometry" (NCTM 2000, p. 96), we realized that we could place the three pigs' houses on a coordinate plane and ask the students to give directions to the wolf as he maneuvered through the village. The development of this mapping activity sparked an idea for exploring both measurement and geometry. Using the tangrams from the previous writing assignment, we asked the children to see how many of the square shapes would be needed to cover the area of the village. We created extension activities that addressed different student ability levels and that included finding the area using other tangram pieces and ultimately varying the village dimensions. This activity involving nonstandard units of measurement allowed the students to see both the need for a standard measure as well as the relationship between the size and shape of different tangram pieces. In knowing and understanding both the NCTM Standards and local expectations, we found ourselves better able to construct and identify the experiences that would enhance our thematic study and further the children's knowledge of significant mathematics.

## A Walk through Our Community

After students read many different versions of "The Three Little Pigs," such as *The Three Little Pigs and the Fox* (Hooks 1989), *Three Little Wolves and the Big Bad Pig* (Trivizas 1993), and *The Three Little Javelinas* (Lowell 1982), the themes of building materials and structures emerged as areas of interest for the students and focal points for continued study. To further explore building materials and structures, we considered looking at structures from both mathematics and social studies perspectives. We began with an investigation of building materials and colors found on the street adjacent to the school. Using a tally sheet, the children recorded data and practiced using and counting tally marks (see fig. 12.5a). On their return to class, the students graphed the data and began to analyze and compare the results. Answering questions using such mathematical language as *most* and *least* as well as such comparative symbols as *greater than* and *less than*, the children generalized their findings.

The children's final activity was to speculate on the meaning of their data within the social context of human behavior. On the basis of their street data, the children completed the prompts "Most houses we saw

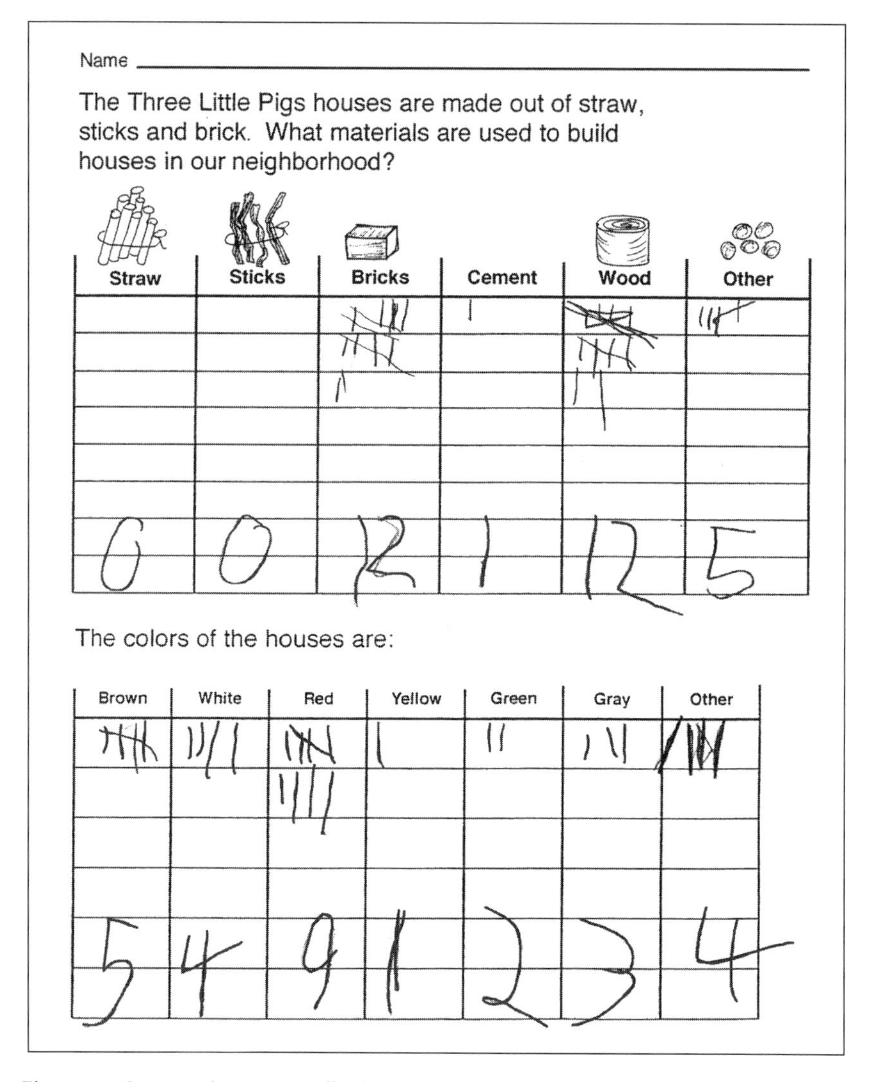

Name ______

The Three Little Pigs houses are made out of straw, sticks and brick. What materials are used to build houses in our neighborhood?

| Straw | Sticks | Bricks | Cement | Wood | Other |
|---|---|---|---|---|---|
| 6 | 0 | 12 | 1 | 12 | 5 |

The colors of the houses are:

| Brown | White | Red | Yellow | Green | Gray | Other |
|---|---|---|---|---|---|---|
| 5 | 4 | 9 | 1 | 2 | 3 | 4 |

Fig. 12.5a. Community-survey tally sheet

were made out of.... I think this is because...." The students' responses reflected thoughtful consideration of data as a tool for interpreting the world around them. Some children commented on the influence of personal preference in their selection of building supplies, whereas others considered the strength of different types of materials. One child reflected on the influence of his data-gathering technique, noting, "When I

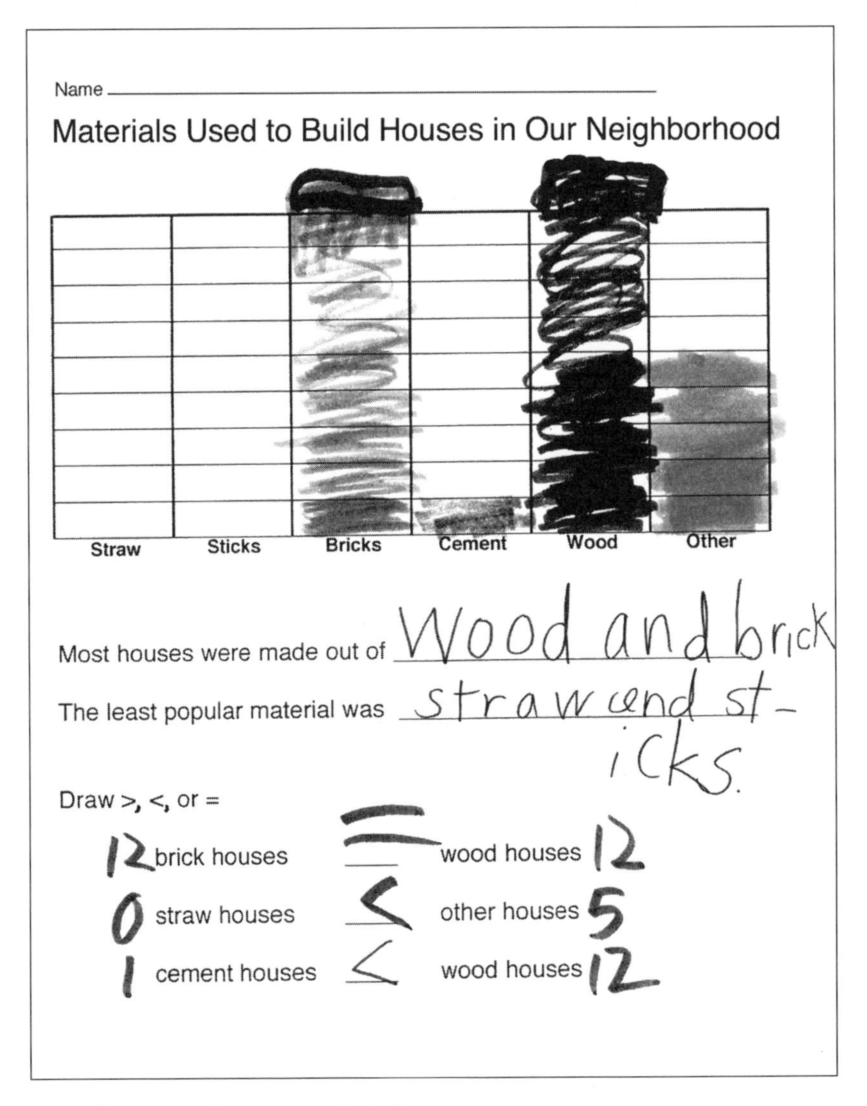
Name

Materials Used to Build Houses in Our Neighborhood

Straw Sticks Bricks Cement Wood Other

Most houses were made out of Wood and brick

The least popular material was straw and st-icks.

Draw >, <, or =

12 brick houses = wood houses 12

0 straw houses < other houses 5

1 cement houses < wood houses 12

Fig. 12.5b. Community-survey analysis sheet

add the chimneys on the houses, that makes the bricks win. But when I count the houses with wood, that makes the wood win." This comment sparked further class discussion about data collection. Throughout this investigation, the students viewed their familiar surroundings through the lens of the mathematician. At the same time, the children were gaining a greater appreciation of their surroundings that had, thus far, gone unnoticed. After this investigation, the children began to report back to the group with observations made in other areas of the community. Furthermore, several parents reported that their children had a greater sense of analytical thinking and an increased interest in statistical surveys.

One of the most exciting aspects of this type of student investigation is the feeling of empowerment that it gives children. Students personally experience and realize that they can take the initiative to conduct similar special-interest surveys. They see the importance and function of data, data analysis, and comparative figures. This meaningful learning experience resembles the work done on higher-order, document-based questions, which seem to be a consistent element in many social studies and English or language arts tests. While helping to prepare students for such examinations, the naturally integrated content in these investigations encourages students to make real connections with their world. Mathematics is brought to life for young students and put to work in the classroom and beyond.

As a follow-up exploration of building structures, we established stations featuring building materials—including marshmallows, straws, toothpicks, gumdrops, building sticks, and traditional blocks—and invited students to take turns in small groups. Later, the children described their experiences with the different materials, including their successes and difficulties. In a focused experiment on three-dimensional shapes, the students used gumdrops and toothpicks to construct a cube and a pyramid. When asked which structure would sustain the greatest weight, the children were quick to say that the cube would be the strongest. Many children commented that it is the same shape as the one in which a traditional home is built. The shrieks that immediately followed the actual weight-bearing experiment—in which the pyramid turned out to be the stronger structure—could be heard in the every corner of the room (see fig. 12.6). One of the most striking benefits of this activity was the heightened interest and attention to block-building. The students' previously flat, simple constructions (see fig. 12.7a) gave way to tall, complex buildings. On reviewing the pictures I had taken to document our year's work, I noticed a significant increase in the participation of girls (see fig. 12.7b). The structured, small-group activities related to our thematic curriculum had, unknowingly, provided the necessary

Fig. 12.6. Students test three-dimensional shapes made with gumdrops and toothpicks.

Fig. 12.7a. Students' block-building before the structure study

Fig. 12.7b. Students' block-building after the structure study

experience for the girls to feel comfortable in the block-building area and thus fostered greater gender equity within the classroom. Surprising benefits seemed to surface everywhere we looked.

## REFLECTIONS ON INTEGRATED CURRICULA

THE CONNECTIONS that children make through integrated curricula are a reflection of the immense preparation on the part of the teacher. Intimately knowing the curriculum; seeking out opportunities in mathematics; and networking, researching, brainstorming, and evaluating an activity is a tremendous undertaking. As seen in the "Three Pigs Ministudy," my collaboration with a colleague was well worth the time and effort of curriculum development. Through our ministudy, unexpected benefits accrued at every turn: increased participation by girls, new opportunities for children of different learning styles, and increased amounts of time on task in mathematics. And even more-far-reaching benefits can be realized. In essence, we teachers are the curriculum weavers that give children a loom through which they can see the beauty of the intertwined threads—an appreciation that in turn encourages them to make deep and lasting connections with their world.

### BIBLIOGRAPHY

Aylesworth, Jim. *The Gingerbread Man.* New York: Scholastic, 1998.

Bartek, Mary. "Hands-On Addition and Subtraction with The Three Pigs." *Teaching Children Mathematics* 4 (October 1997): 68–70.

Bell, Nanci. *Visualizing and Verbalizing for Language Comprehension and Thinking.* Paso Robles, Calif.: Academy of Reading Publications, 1991.

Hankes, Judith E. "An Alternative to Basic-Skills Remediation." *Teaching Children Mathematics* 2 (April 1996): 452–57.

Hong, Lily Toy. *Two of Everything.* Morton Grove, Ill.: Albert Whitman & Co., 1993.

Hooks, William H. *The Three Little Pigs and the Fox.* New York: Macmillan Publishing Co., 1989.

Johnston, Tony. *Farmer Mack Measures His Pig.* New York: Harper & Row, 1986.

Lowell, Susan. *The Three Little Javelinas.* New York: Scholastic, 1992.

Maccarone, Grace. *Three Pigs, One Wolf, and Seven Magic Shapes.* Activities by Marilyn Burns. New York: Scholastic, 1997.

Marshall, James. *The Three Little Pigs.* New York: Dial Books for Young Readers, 1989.

Mid-continent Research for Education and Learning (McREL). *Comprehensive School Mathematics Program,* CSMP/21. Aurora, Colo.: McREL, 1991.

National Council of Teachers of Mathematics (NCTM). *Professional Standards for Teaching Mathematics.* Reston, Va.: National Council of Teachers of Mathematics, 1991.

———. *Principles and Standards for School Mathematics.* Reston, Va.: NCTM, 2000.

Scieszka, John. *The True Story of the Three Little Pigs.* New York: Puffin, 1996.

Trivizas, Eugene. *Three Little Wolves and the Big Bad Pig.* New York: Scholastic, 1993.

13

# Connecting through Intregrated Mathematics

Nancy G. Nagel

"HOW DO you know that the little kids want more playground equipment?" "What are 'square feet'?" "How do we find out how much grass seed we need to buy for the soccer field?" These questions arose during discussions in a fifth-grade class taught by Monica, one of my preservice teaching students at Lewis & Clark College in Portland, Oregon. As her advisor and professor, I had the opportunity to observe her teaching internship. To help the students answer these questions, she created a mathematical unit that integrated geometry, computation, estimation, data analysis, and measurement through the context of their real-world problem. Many connections among these mathematical topics, between mathematics and other curricular areas, and between mathematics and the real world emerged throughout this problem-solving unit.

The real-world problem came from the students' concerns about their playground. Day after day, they returned from recess with muddy shoes from the playing fields and complaints about graffiti and ripped basketball nets. Monica took these concerns seriously and constructed a unit that integrated mathematics with other curricular areas as well as used mathematics to connect the students with their local community. *Principles and Standards for School Mathematics* (NCTM 2000, p. 182) states, "The goal of school mathematics should be for all students to become increasingly able and willing to engage with and solve problems." Studying their school's playground gave Monica's students a meaningful problem that they were eager to learn to solve.

## REAL-WORLD PROBLEM SOLVING AND MEANINGFUL MATHEMATICS

REAL-WORLD problem solving is a philosophy of teaching and learning in which students work together to solve a problem or issue of concern to them, their teacher, and their local community. In problem solving, students research a topic and develop and implement a "best possible solution" when appropriate. Prior to their research, students see no apparent solution. To obtain current data, students often interview local experts in the field and use technology to collect relevant data. The goal of learning shifts from memorizing facts to obtaining and using knowledge, skills, and understandings to solve a problem. The term *real world* is not meant to delineate learning within or outside the school but rather to emphasize the essence of students' ownership of the problem, the solution, and the connection with the local community (Nagel 1996).

Throughout the playground problem, Monica's students learned and used mathematics to collect, organize, interpret, and display data and to make informed decisions that led to solutions. Monica integrated different topics in mathematics, and these connections helped the students obtain and create accurate and meaningful information with which to analyze the problem.

### Planning for the Playground Unit

Prior to the unit, the class members discussed their concerns about their playground. Monica asked such questions as "Are you interested in improving the playground?" and "What are some improvements that could be made?" The students' responses to the latter question included planting grass seed, adding new equipment, painting over graffiti, and planting flowers around the edge of the fields. These responses helped Monica identify the areas of mathematics to integrate throughout the unit.

Monica began by rereading and reviewing *Principles and Standards for School Mathematics* (NCTM 2000) and its discussion of the four major content standards (Number and Operations, Geometry, Measurement, and Data Analysis and Probability) essential to her unit. She wanted to ensure that she incorporated the major goals and expectations of each Standard for students in grades 3–5. She planned her unit so that number-and-operations goals would build a foundation for understanding.

Students would employ number and operations to compute and use addition, subtraction, multiplication, division, and estimation to measure the playground (fig. 13.1), develop a budget to buy new playground equipment, and create surveys and graphs. In geometry, students would use visualization, geometric reasoning, and geometric modeling to conceptualize and solve the problems associated with the playground. Monica would have students draw maps of the playground and construct models of the proposed playground. Both components of the Measurement Standard (NCTM 2000, p. 44) would come into play as students determined and reported measurable attributes of the playground and selected and used appropriate measurement techniques, tools, formulas, and units for determining the length, width, perimeter, and area of the playground.

Fig. 13.1. Students measure bald spots in the field to determine the amount of grass seed needed.

Monica's students had already studied aspects of those three Standards and would now apply their knowledge, skills, and understandings. Skills from the fourth Standard, Data Analysis, would be crucial for (*a*) formulating survey questions; (*b*) collecting, organizing, and displaying data; (*c*) constructing charts and graphs; and (*d*) making and evaluating inferences and predictions. Although the students had worked with charts in previous lessons, in this unit Monica would need to introduce the concepts of creating surveys and graphing the results.

In addition to the Content Standards, Monica would use each of NCTM's five Process Standards during the unit. With the entire unit based on a real-world problem, she could touch on all four goals of the Problem Solving Standard (NCTM 2000, p. 52). Her students would (1) build new mathematical knowledge, (2) solve mathematical problems as they arose, (3) apply and adapt a variety of problem-solving strategies, and (4) discuss and reflect on the mathematical problem-solving process—all within the context of improving their playground. Regarding the Reasoning and Proof Standard, students would develop possible solutions and justify and prove their conjectures.

The unit also incorporated the goals of the Communication Standard in that students would share their mathematical thinking (e.g., explanations of survey results) and analyze and evaluate their own work and that of others as they used mathematical language to express mathematical ideas. Because the unit highlights the relationships within mathematics and between mathematics and other areas of learning, the Connections Standard was a major focus. For example, students would be measuring the playground, building models of the playground to scale, sharing survey findings, and presenting their problem with possible solutions. All these activities draw from the connections within mathematics and the application of mathematical ideas to improve the playground. When students create tables, graphs, or drawings, they are representing their mathematical ideas and data to clarify and show their understanding of the physical, social, and mathematical aspects of their problem while also communicating their findings.

Next, Monica considered different projects that would allow students to work with these Standards and help them find a solution. She researched the NCTM Standards for students at the fifth-grade level so that she could teach appropriate skills and concepts. At the same time, Monica wanted to leave the projects and activities open-ended enough for her students to make choices as they needed data or sought information throughout the problem-solving process. For example, Monica wanted her students to know how to construct bar, circle, and line graphs, so she planned lessons to introduce data collection and the construction of graphs. In a sense, this design is a "backward" one (Wiggins and McTighe 1998), where the teacher looks at the desired learning goals and plans learning experiences leading to these results.

In addition to reviewing the mathematics Standards, Monica planned lessons to allow her students to conduct research on the Internet. Monica worked with small groups to select key words or phrases for Web searches, for example, *school playground* or *graffiti removal.* From these searches, the students could find information on playground equipment or on local agencies that could help them remove the graffiti. Other lessons focused on conducting telephone interviews with experts in local agencies and on organizing and presenting speeches. These preparatory lesson plans helped Monica prepare the students to research their playground problem and find solutions.

## The Playground Unit

In this unit, the mathematics (i.e., data analysis, measurement, geometry, and number and operations) was students' primary tool for conducting research and organizing and presenting their findings. Instead of learning each separate area of mathematics in isolation, the students pulled as needed from multiple mathematical topics. For instance, following a lesson on measurement and square feet, the students determined the size of the playing field that was bald and frequently muddy. Two students, Micah and Stacia, discussed how multiplying the sides of a rectangle translated to measuring the length and width of the field. Then they multiplied these two numbers, or sides, to find the square footage. They said, "See, we can draw a picture of our field and write what we measure on the side, just like we did on the overhead. Then we will have square feet." These two students not only learned the formula for determining square feet but also learned when and how to apply mathematics to real-life situations. Here we can see the power of relevance in learning mathematics. In the context of the playground, measurement provided natural links between experiential learning and the world of number (Pope 1994).

For Monica, a major struggle in her teaching was deciding when to step in and work with the students on a specific skill or understanding and when to step back and wait for the students to need this information in their work. When working with data analysis, she found that the students seemed to understand graphing at a deeper level when they constructed graphs from their own questions and data. In this unit, Monica used samples of students' data to present several short lessons on types of graphs and representations of data. Then she asked the students to create their own graphs. The impetus for the next component of the unit came when one student, Jamile, asked, "How will we know what other kids want on the playground?" In response, the students decided to survey other students in the school. Monica asked, "What will you ask them, and how will you know what all the answers mean?" The students decided that they needed to create a survey—in this instance, a list of four questions—and graph the students' responses.

The graphs revealed that more students wanted another basketball

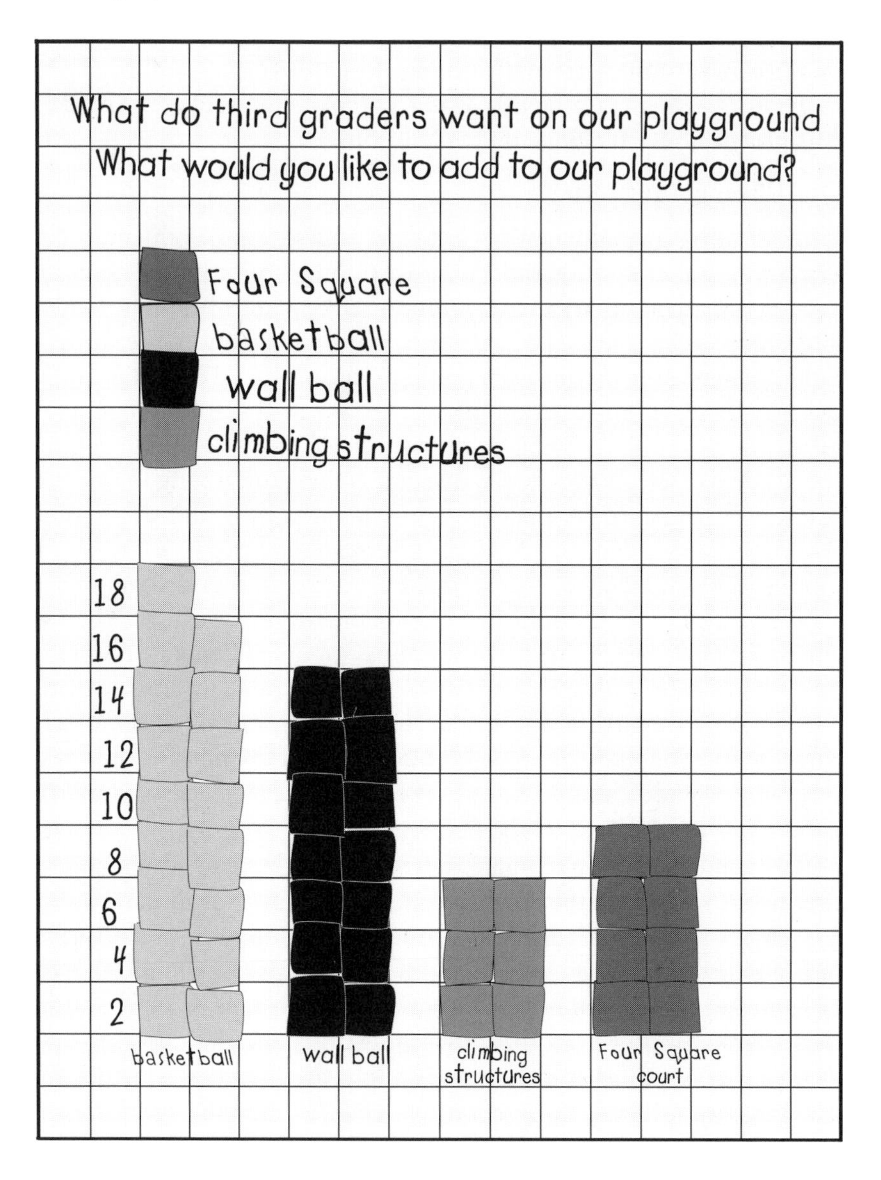


Fig. 13.2a. Graph of students' survey responses

court than wanted additional wall-ball space. Most students created bar graphs (see figs. 13.2a and 13.2b), although several groups used circle graphs to display the survey results. Monica understood the central importance of graphs to statistics (Clark 1994); therefore, she linked the students' survey questions with the statistics

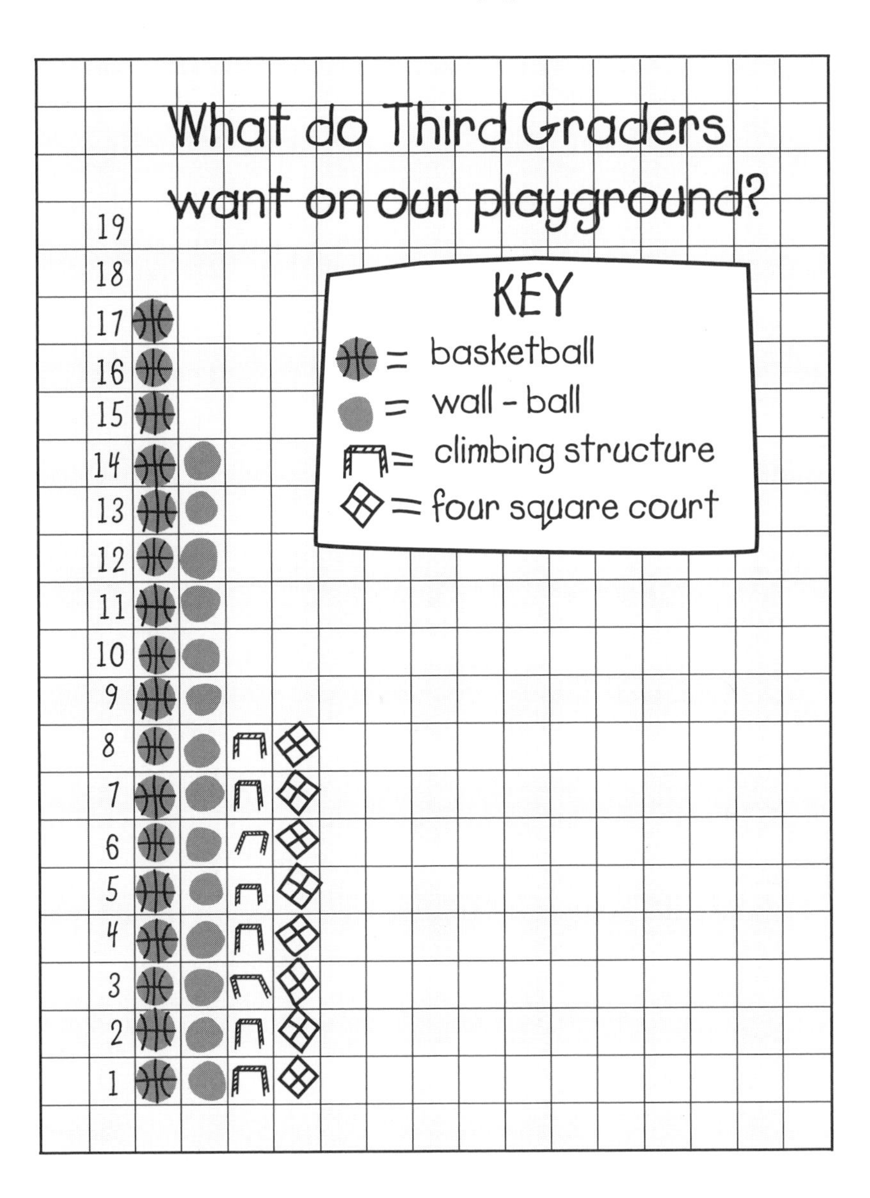


Fig. 13.2b. Graph of students' survey responses

they had gathered and helped them accurately represent these statistics through graphing.

Throughout the unit, Monica found that small-group work allowed the students to process information. The interactions among the students fostered discussions about mathematical ideas and questions. When a group of five students tried to figure out how much grass seed would be needed to plant the soccer field, they tested each other's strategies for determining square footage and applied the formula found on the grass-seed bag. Next they drew a table to show their calculations and proposal for the amount of grass seed to purchase. Then they presented the table to their classmates, who agreed with their suggested purchase. The group interaction allowed for testing formulas and checking both strategies and computations.

## Assessment of Students' Learning

Monica decided to use multiple assessment formats connected with the unit's content to authentically assess her students' learning. She wanted formats that would enable her to account for students' learning within mathematics as well as in their application of mathematics (Webb 1992). With these goals in mind, she selected the formats of interviews, observations, portfolios, and student-developed scoring guides. In the portfolio, each student placed the following:

- A copy of the group's survey questions
- A sample bar graph
- A map of the playground, including dimensions and area
- A budget sheet listing income and expenditures for the playground improvements
- An outline of a speech to present to the parent club
- A persuasive essay presenting the student's suggestions for improving the playground
- A self-assessment of learning gains from the problem-solving project

For her observations of students' work, Monica developed a form with a list of students' names and space for notes (see fig. 13.3).

While the students measured the playground, Monica recorded information about each student's (*a*) use of appropriate tools and units, and (*b*) computational skills when measuring, adding feet and yards, and converting within the unit of yards. The variety of assessment formats led to a more comprehensive portrayal of students' learning.

| MEASUREMENT OBSERVATION | | |
|---|---|---|
| NAMES | DATE | NOTES |
| Ada | | |
| Anthony | | |
| Antoinette | | |
| April | | |
| Betsy | | |
| Carolina | | |
| Erin | | |
| Jaina | | |
| Jamile | | |
| Juan | | |
| Katrina | | |
| Kris | | |
| Lauren | | |
| Maile | | |
| Marc | | |
| Maria | | |
| Mary Gage | | |
| Megan | | |
| Micah | | |
| Roberto | | |
| Scott | | |
| Stacia | | |
| Taneesha | | |
| Tobie | | |

Fig. 13.3. Form for teacher's notes from observation of student work

Monica worked with her students to develop several scoring guides for their work; doing so helped the students understand the expectations as well as enabled them to offer ideas on what attributes make a graph, presentation, survey, or final project outstanding. For example, following an initial lesson on creating bar graphs—but prior to the actual creation of bar graphs to represent data from the surveys—Monica and the students determined criteria for three different levels of progress in learning graphing. At level 1, the students would have little or no understanding of graphing; at level 2, the students would have some knowledge of graphing; and at level 3, the students would be able to graph with accurate explanations and interpretations of data (see fig. 13.4).

| | RATING | | |
|---|---|---|---|
| CRITERIA | 1 | 2 | 3 |
| Did I include all important information? | Missing all or most | Has most of the information | Complete information included |
| Did I make a key to tell what the numbers mean? | No key | Key with one point | Key that explains value of bars |
| Is the information accurate? | Inaccurate | Mostly accurate | All information is accurate |

Fig. 13.4. Scoring guide for students' self-evaluation of their graphs

The multiple forms of assessment fostered both formative and summative assessments of students' learning. Monica noted the students' progress and areas in which they needed more work. Instead of waiting until the end of the unit, she gave additional instruction as needed to assist students in mastering the skills, concepts, and knowledge necessary to use mathematics in their problem solving. Her assessments generated feedback on students' learning and enabled Monica to make informed decisions about her teaching. Her teaching and assessment were connected and interrelated, and each influenced the other.

## Mathematics and Connections

As the fifth-grade students neared completion of their six-week study of the playground, they planned a community day to clean up the school grounds, paint over graffiti, plant grass seed and small plants, and install new playground equipment. Following their presentation to the parent club, the students were awarded $800 for playground equipment and supplies. A local store sold them grass seed, plants, and paint at a reduced cost. The students posted signs around their neighborhood, inviting community members to participate in the community day. More than one hundred people arrived to help with the project. The following week, the city's commissioner of parks came to the school and rededicated the playground.

As Perrone (1994) points out, "Our students need to be able to use knowledge, not just know about things. Understanding is about making connections among and between things, about deep and not surface knowledge, and about greater complexity, not simplicity." These young children studied mathematics and learned about and used number and operations, geometry, measurement, and data analysis. At the same time, the mathematics was connected communication through writing, telephone interviews, and speeches. The students used fine-arts skills to create posters and models of the playground. The social sciences provided background knowledge for surveying and examining different perspectives of issues. Science knowledge helped students select the grass seed appropriate for playground use and for the school's soil conditions. Within the context of a real-world, meaningful problem, the students learned across multiple curricular areas and mathematical topics. Because of the many connections, the mathematics was integrated, relevant, and meaningful.

## REFERENCES

Clark, Megan. "Teaching Statistics." In *Mathematics Education: A Handbook for Teachers,* edited by Jim Neyland, pp. 79–87. Wellington, New Zealand: Wellington College of Education, 1994.

Nagel, Nancy G. *Learning through Real-World Problem Solving: The Power of Integrative Teaching.* Thousand Oaks, Calif.: Corwin Press, 1996.

National Council of Teachers of Mathematics (NCTM). *Principles and Standards for School Mathematics.* Reston, Va.: NCTM, 2000.

Perrone, Vita. "How to Engage Students in Learning." *Educational Leadership* 51 (February 1994): 11–13.

Pope, Lindsay. "Teaching Measurement." In *Mathematics Education: A Handbook for Teachers,* edited by Jim Neyland, pp. 100–105. Wellington, New Zealand: Wellington College of Education, 1994.

Webb, Norman. "Assessment of Students' Knowledge of Mathematics: Steps toward a Theory." In *Handbook of Research on Mathematics Teaching and Learning,* edited by Douglas A. Grouws, pp. 661–86. New York: Macmillan, 1992.

Wiggins, Grant, and Jay McTighe. *Understanding by Design.* Alexandria, Va.: Association for Supervision and Curriculum Development, 1998.

14

# Using Children's Literature to Support Early Childhood Mathematics Education

Virginia L. Keen

Until recently, teacher-preparation programs did not devote much attention to children's academic learning prior to kindergarten; rather, they focused predominantly on nonacademic content (e.g., nutrition and family issues). However, increased attention to building the foundations of mathematical thinking prior to kindergarten raises new concerns for mathematics teacher educators. In *Principles and Standards for School Mathematics* (*Principles and Standards*) (NCTM 2000), the National Council of Teachers of Mathematics extended its guidelines for high-quality mathematics teaching and learning to the preschool level. In addition, recent changes in state teacher-licensure programs emphasize preparing prospective preschool-level teachers at the same substantive level as prospective elementary school–level teachers (e.g., Ohio Department of Education's [1990] early childhood licensure for teachers of prekindergarten through grade 3). As a result, colleges and universities are designing courses in mathematics content and pedagogy specifically related to early childhood education. Teacher educators must ask themselves how they can improve what is included in mathematics methods courses as well as how they teach these courses and how they support the work of practicing teachers.

One important tool for improving early childhood mathematics education is integrating children's literature into methods courses and subsequently into classroom mathematics instruction. Children's literature can be a powerful component of any content-area study. For mathematics instruction, integrating children's literature has many benefits. Griffiths and Clyne (1991) note that "Using books, stories, and rhymes to stimulate thinking about mathematics and to develop and reinforce mathematical concepts enhances children's understanding of mathematics, promotes their enjoyment of the subject, and develops their conception of mathematics as an integral part of human knowledge." (p. 9)

In addition, Ross (1994) identifies nonfiction books as valuable tools for "increasing vocabulary, broadening experiences, helping children see connections among school subjects, letting them become problem solvers and researchers, and luring them into learning content area information" (p. 35). She observes that "[w]hat might seem cold, dull, and uninspiring in a textbook often comes to life when presented by an author who is impassioned about the subject" (p. 13).

In *Feisty Females: Inspiring Girls to Think Mathematically*, Karp and others (1998) describe another strong motivation for integrating children's literature. For two years, they studied a mathematics curriculum that "featured rich stories of females who were risk-takers, problem solvers, and who used relationships with others to tackle difficult situations" (p. 4). They found that "[i]f teachers wish to develop the underlying traits that strengthen problem-solving abilities of females, they need to look at ways to introduce female characters in literature and real people with hardy personalities into their instruction. Clearly, there is a necessity to include female role models in classrooms so that teachers meet the needs of all children." (p. 138)

These multiple purposes as well as the research findings give ample reasons for integrating children's literature and mathematics.

## PROFESSIONAL SUPPORT FOR INTEGRATING LITERATURE

SEVERAL PROFESSIONAL organizations encourage integrating mathematics and literature. *Principles and Standards* (NCTM 2000, p.118) notes that children's literature can set the context for both student-generated and teacher-posed problems. Furthermore, the five NCTM Process Standards—Problem Solving, Reasoning and Proof, Communication, Connections, and Representation—can be addressed by thoughtfully integrating children's literature into, and using associated inquiry-based teaching strategies with, mathematics. The connection between children's literature and NCTM's Standards is not new, however. In 1992, David

Whitin assembled a rich collection of stories that showed how children's literature could enhance Standards-based mathematics education, particularly the learning called for in *Curriculum and Evaluation Standards for School Mathematics* (NCTM 1989). Whitin's collection is a still useful and applicable guide.

The National Association for the Education of Young Children (NAEYC) calls for integration across traditional subject-matter divisions "to help children make meaningful connections and provide opportunities for rich conceptual development" (NAEYC 1997, p. 41). The natural way that students learn is more accurately reflected in topics that are integrated, "not compartmentalized or divided into artificial subject-matter distinctions" (p. 66). According to the National Council of Teachers of English (NCTE) in its *Guidelines for the Preparation of Teachers of English Language Arts* (NCTE 1996), "Instruction that calls for subject matter across subject lines is increasingly seen by the profession as important, both because an integrated curriculum will increase richness and because it will give greater meaning to each of the disciplines" (p. 24). As part of its "Five Core Propositions" of accomplished teaching, the National Board for Professional Teaching Standards asserts that accomplished teachers understand their subject and how to link it with other disciplines (NBPTS 2002). These various disciplinary and certification standards show why preservice and in-service teachers must be able to make good instructional decisions about integration (e.g., linking children's literature with mathematics instruction).

At the same time that integrating literature increases young students' enjoyment of mathematics, it also helps prospective teachers who have math anxiety. Those preservice teachers can approach mathematics from a language arts perspective, an area in which many are typically more comfortable. "Approaching math this way [through literature set in everyday activities] demystifies it and shows that 'doing' math is really an activity that is part of everyone's life, an activity that we do all day long" (Freeman and Person 1998, p. 41). By learning how to use literature as an instructional tool, future teachers gain confidence in their own ability to make sense of mathematical ideas. But just as learning mathematics comes from *doing* mathematics, so does learning to choose and use children's literature as a vehicle for mathematics learning result from actually preparing such lessons.

In general, past preservice teachers were urged, but not required, to use children's literature in mathematics instruction. But encouraging the use of children's literature in lessons is not the same as expecting thoughtful choices of children's literature in teachers' planning. Prospective teachers hone the tools for the task only when they practice and learn how to make good selections and how to effectively incorporate, rather than just add, literature. Too often, novice teachers have not had adequate experience in transforming a good story into a vehicle for mathematics instruction.

Mathematics teacher educators who are inexperienced with integrating children's literature in mathematics learning at the preschool level should experience the planning process. To guide teacher educators' own learning as well as their teaching of preservice teachers, some suggestions regarding this instructional methodology are given below.

## CREATING A LIBRARY

TO SKILLFULLY include children's literature in mathematics instruction, one must first become familiar with and collect children's literature that can be used to enhance mathematics learning. Many children's books can be related to mathematics in some way, but not all are equally well suited. Chatton and Collins (1999, p. 125) note the following:

> As we work with our students, our curriculum, and good books, possibilities emerge at every turn. We need to consider which links are natural, giving rise to questions that reveal differences in how people in different disciplines might view the world. We need to invite students to try new ways of thinking about the world and to keep ourselves from far-fetching simply because we notice overlapping terms or concepts.

One way to wisely select books is to learn from educators with considerable experience in integrating literature. For example, I asked a retired mathematics supervisor to identify books that she knew worked well in elementary school mathematics lessons. Her recommendations seeded my library. One of her recommendations—a large and inexpensive listing of children's literature that includes titles, ISBN numbers, authors' names, grade-level approximations, and mathematics topics (Huellmantel 1999)—yielded numerous resources.

Other possible library additions can be found in (a) the previously mentioned Whitin (1992) article; (b) materials designed to support connecting mathematics and literature (e.g., *Books You Can Count On: Linking Mathematics and Literature* [Griffiths and Clyne 1991] and *Math and Literature (K–3)* [Burns 1992]); and (c) the "Links to Literature" department in issues of *Teaching Children Mathematics* (*TCM*). For example, in one *TCM* article, "Selecting Books in Spanish to Teach Mathematics" (Jacobs, Bennett, and Bullock 2000), the authors recommend literature related to several topics and give sound advice that holds true for mathematics instruction in any language.

Beneficial books include stories with no words (e.g., *Changes, Changes*, by Pat Hutchins [1987]) and stories with many possible subject-matter connections (e.g., *One Watermelon Seed*, by Celia Barker Lottridge [1990], for mathematics, language arts, and science; or *Sir Cumference and the First Round Table*, by Cindy Neuschwander [1997], for mathematics, language arts, and social studies). Several books written by Stuart Murphy for the Math Start series are also helpful. The cover of each Math Start book features a small triangle that shows the highlighted mathematical concept or operation. Each book also includes a "For Adults and Kids" section with suggestions for questions, additional activities, and related children's books (see this chapter's Bibliography for two Stuart Murphy books). Other authors' books often have similar supplemental information. These added suggestions make such books especially valuable for inexperienced teachers and for use in "books in a bag" collections that students take home for family reading and related activities.

When selecting books, remember your purposes and motivation for integration. Do not merely choose a title related to the next mathematics concept covered in the textbook. Make sure that the book can make the concept more meaningful.

When acquiring books, note the mathematical concepts that each book deals with effectively. Then record your notes in your state's curriculum guide. You will have a convenient reference for linking literature and mathematics content as well as a tool to share with preservice teachers—a tool that shows how to organize literature in support of mathematics performance objectives. Your curriculum-guide notes can serve as a quick and easy reference, but a database that includes this information might be even more helpful.

## Developing Lesson Plans—a Teacher Education Example

In my course for preservice teachers, I ask the students to create at least one classroom lesson plan that incorporates children's literature. The students may use ideas gleaned from *TCM* (which includes articles on how teachers have used specific books as well as helpful reviews of children's literature) as jumping-off points for integrating literature into what

Baroody (2000) describes as "purposeful" mathematics instruction. I also encourage my preservice teaching students to discuss how to employ children's literature and other books in lessons during their student teaching.

However, suggesting the use of children's literature, recommending books, and teaching students how to create lesson plans are not sufficient steps for preparation. We know that experience in using manipulatives is not enough to ensure that teachers will use these tools appropriately. Similarly, laying the foundation for good lesson planning with literature must include discussions of appropriate use and thoughtful integration.

One strategy that I use is modeling the placement of literature in mathematics content and methods lessons. One lesson that I model uses part of *Grandfather Tang's Story* by Ann Tompert (1990) to initiate work with the concept of area using tangrams. In this Chinese folktale, Grandfather Tang describes two fox fairies that transform themselves into different animal shapes, all made from two sets of tangrams. In the lesson, the teacher asks the students to try to recreate the animal shapes at their tables and to learn to recognize the size properties of each tangram piece. Following the story, the class determines exactly what area relationships exist among the pieces. Then the students go a step further and identify the story's "math moral"— a general mathematics principle that can be abstracted (Friedman 1997). Friedman suggests that "[b]y encouraging children to look for, discover, discuss, and extend math morals, teachers can help them understand important mathematical principles and gain experience in generalization and abstraction, which are important higher order thinking skills" (p. 33). In *Grandfather Tang's Story*, the math moral might be stated as "Given a fixed set of two-dimensional shapes, their total area is equivalent no matter how the pieces are rearranged."

That math moral leads naturally to a discussion of area, fractions, and the effect of changing units. Mathematical concepts and questions to investigate include the following:

- *Fixed area.* What can you say about the areas of the two fox fairies? How do you know that they have the same area? (Because they can be made from the same set of tangrams using all pieces, they hence have equivalent area.)
- *Importance of defining a unit.* If a small triangle is one square unit (can it be one square unit if it isn't square?), what are the areas of the other pieces? If a big triangle is one square unit, what are the areas of the other pieces?

- *Proportional reasoning.* How can you use what you know about the areas when a small triangle is one square unit to find the areas of the pieces when the large triangle is one square unit?
- *Mathematics pedagogy.* The figures in the story have the shapes outlined in their pictures (see fig. 14.1a). What if the helping lines were excluded (see fig. 14.1b)?

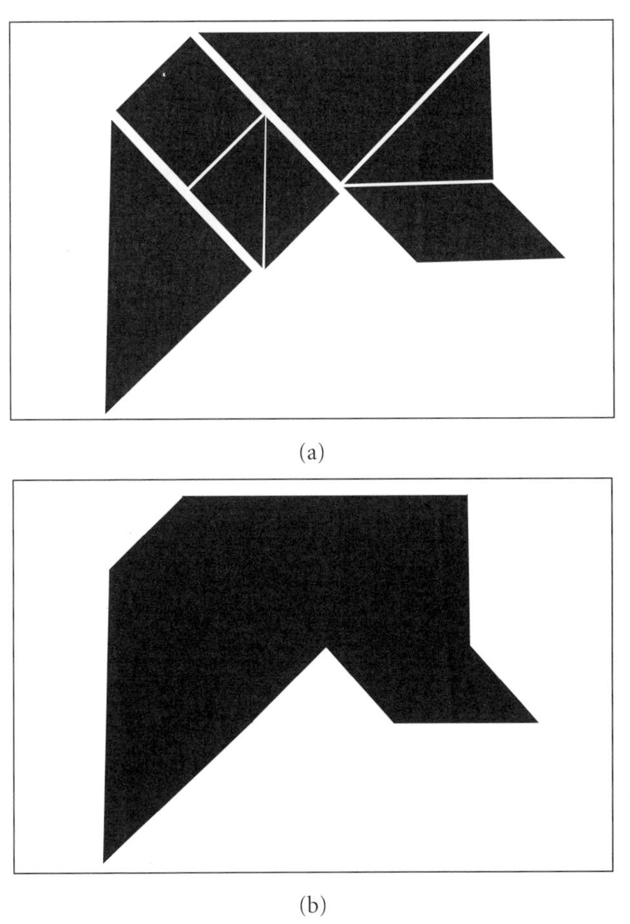


Fig. 14.1. Tangram helmet shown (a) with help lines and (b) without help lines

Thus, a particular piece of children's literature can serve as an appropriate vehicle to study many different mathematical concepts at different grade levels. The teacher's role is to guide the discussion of the particular concepts for the specific lesson. A book like *Grandfather Tang's Story* can also serve as a tool for cultural learning as well as for a life lesson, illustrated through the nearly tragic experiences of the fox fairies.

Other materials can be used to support the use of *Grandfather Tang's Story* and tangrams to explore additional geometry content (Ernst and Ernst 1990; Russell and Bologna 1987; Dunkels 1990). This exploration can be extended to other geometry concepts and the use of other children's books (Harris 1998; Thiessen and Matthias 1989). For younger children needing experience with shapes, a wealth of children's books and resources include (a) Web sites, (b) the *TCM* Focus Issue *Geometry and Geometric Thinking* (1999), and (c) such books as *Shapes, Shapes, Shapes* (Hoban 1986) and *Sea Shapes* (MacDonald 1998).

With the variety of books available, the teacher needs to make an informed choice after looking at many books and having the students give feedback. Such books as Tana Hoban's *Shapes, Shapes, Shapes* (1986) are well liked because they include photographs that make explicit the real-world connections. However, many books with colorful drawings are equally appealing and helpful for exploring mathematics.

Stories can be integrated at different points in the lesson. For example, when a story introduces a mathematical concept at the opening of a lesson, students can learn vocabulary and begin thinking about how the characters in the story solve the problems with which they are faced. Introducing the story first can initiate whole-class discussions, small-group experimentation, or individual study. Some books lend themselves to examination over a whole class period with the teacher guiding the discussion as the story unfolds. In some instances, a book can serve as a daily prompt to explore new ideas over several class periods.

The same books can play different roles at different grade levels. For example, in a kindergarten class, *The Doorbell Rang* by Pat Hutchins (1989) can be told by using a felt board and felt cookies and can involve the entire class over the whole class period. With a third-grade class, the book can be used to introduce the idea of partitive division.

A story presented at the end of a class period can help tie together ideas. The story can offer closure or foreshadow mathematical ideas to come. The story might be revisited briefly at the start of the next lesson. In all instances, the students' understanding should be assessed at the end of the lesson so that the teacher can make informed instructional decisions about upcoming lessons.

No matter where a story is placed in the structure of the lesson, two steps are essential. First, the teacher must read the story all the way through shortly before reading it to the class. This simple but important task ensures that the teacher can guide the discussion toward learning goals. In one instance that I observed, the teacher had examined a book and had seen material suitable for introducing a particular concept. However, she had not read through the whole story, and her students became confused when the story introduced an unanticipated concept.

Second, teachers should create a list of possible questions to ask during or after the reading to help focus students' attention on the mathematics. Creating that list will help the teacher, even if many of the generated questions are not actually used in the class discussion. Tailored questions can also be used to assess students' understanding and use of vocabulary.

## SHARING LESSON PLANS

TO GAIN experience with these aspects of lesson preparation, each of my methods students creates a mathematics lesson that integrates children's literature. The students prepare the plan, present a synopsis to the class, and share information about the literature incorporated in the lesson. The sharing includes showing the story's illustrations and describing how the illustrations relate to or emphasize the mathematical ideas addressed. The students also read aloud portions of the story to illustrate the mathematical content, the reading level, and the author's writing style. Figure 14.2 shows a transparency created in October 2000 by one of my students, Mindy Lewis, to explain her lesson incorporating the story *Esio Trot* by Roald Dahl (1992).

In each methods class, prospective teachers are introduced to the literature used in their classmates' lessons and become familiar with various lesson designs. The lesson plans are copied for each member of the class. When these students enter the field, the lesson plans serve as a ready source of ideas. Even when they are teaching grade levels different from those addressed by the prepared lessons, the prospective teachers can use the same literature for their classes by changing the level of questioning, the pace of reading, and other aspects of the teaching and learning environment.

## ENGAGING ALL STUDENTS

Prospective preschool-level teachers who previously were not expected to gain the subject-specific knowledge required of elementary school–level teachers are now in teacher education programs. Many students majoring in early childhood education assume that the mathematics component of their program will not be demanding. Those of us designing the required courses for these students must consider

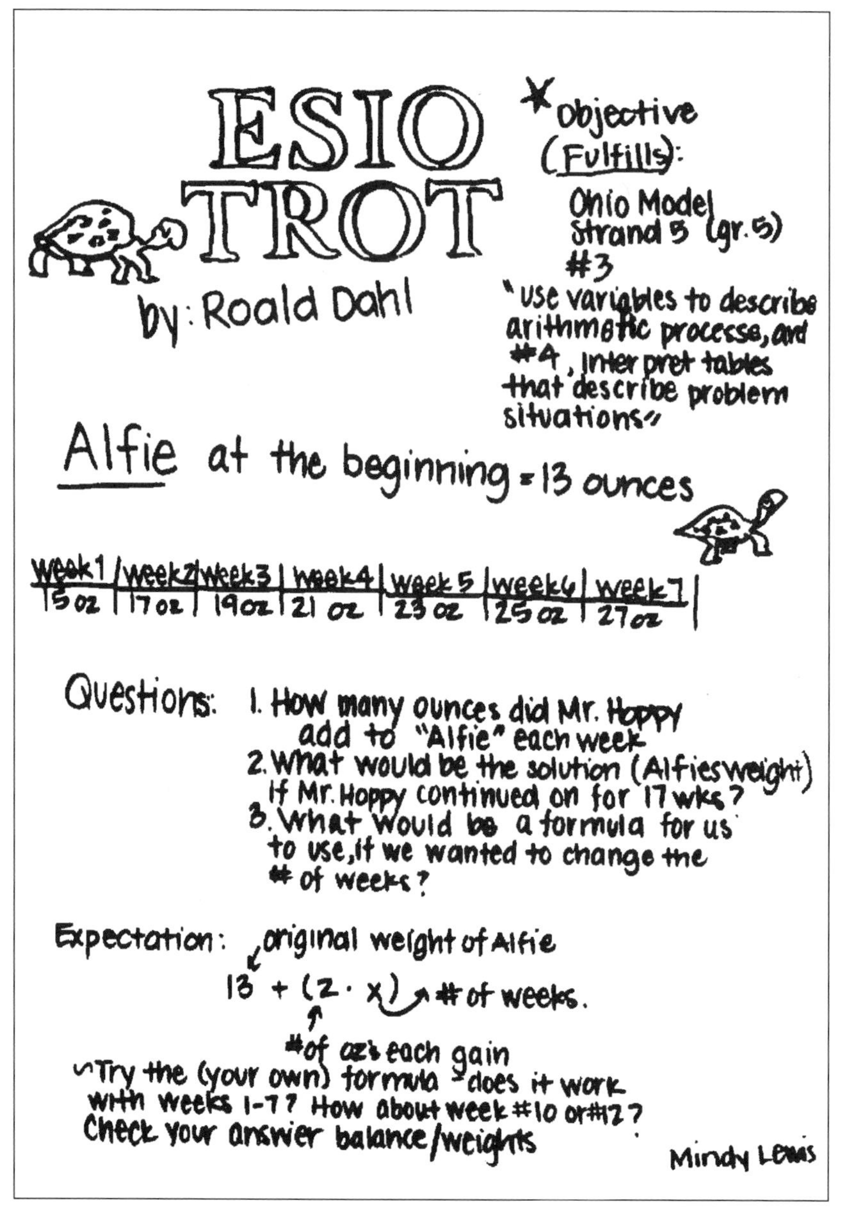


Fig. 14.2. A transparency created by preservice teacher Mindy Lewis to explain her mathematics lesson plan based on *Esio Trot* by Roald Dahl (1992). Used with permission of Mindy Lewis.

these preconceptions. Integrating children's literature not only creates a comfortable setting but also offers a powerful model for mathematical content and pedagogical learning. Post and his colleagues reason that "[i]f we teach in a manner that isolates the various disciplines, then students learn that the disciplines are separate, unconnected, and discontinuous. If we teach them integrated approaches, that is what students will learn" (1997, p. iv). Mathematics methods students who have developed lesson plans that integrate children's literature report that this experience is an important one for them. As a result of the confidence gained in planning with literature, the students indicate that they are more open to, and able to see possibilities for, integrating other subjects into their mathematics teaching.

Clearly, integrating literature into mathematics teacher preparation is productive. Such integration can be a major building block in methods courses for early childhood teachers. Furthermore, cross-disciplinary cooperation from other content and methods instructors offers rich opportunities for preservice teachers to learn about mathematics and teaching as well as the role of children's literature in educational development. Incorporating appropriate children's books connects children's mathematics learning with their learning in other disciplines.

Integrating other subjects with mathematics can show preservice teachers how to foster opportunities for children to learn—but it is very important that the mathematics not be lost. Stories that are pleasant or captivating but lack the substance to develop deeper mathematical understanding are of little value. Fortunately, an increasing number of children's books offer rich contexts for asking important mathematical questions of students of any age. By examining and using appropriate children's literature, mathematics teacher educators and mathematics teachers can more effectively achieve their goals of meaningful mathematics teaching and learning.

## BIBLIOGRAPHY


Baroody, Arthur J. "Does Mathematics Instruction for Three- to Five-Year-Olds Really Make Sense?" *Young Children 55* (4) (July 2000): 61–67.

Burns, Marilyn. *Math and Literature (K–3).* Sausalito, Calif.: Math Solutions Publications, 1992.

Chatton, Barbara, and N. Lynne Decker Collins. *Blurring the Edges: Integrated Curriculum through Writing and Children's Literature.* Portsmouth, N.H.: Heinemann, 1999.

Dahl, Roald. *Esio Trot.* New York: Puffin Books, 1992.

Dunkels, Andrejs. "Making and Exploring Tangrams." *Arithmetic Teacher* 37 (February 1990): 38–42.

Ernst, Lisa Cambell and Lee Ernst. *The Tangram Magician.* New York: Harry N. Abrams, Publishers, 1990.

Freeman, Evelyn B., and Diane Goetz Person. *Connecting Informational Children's Books with Content Area Learning.* Needham Heights, Mass.: Allyn & Bacon, 1998.

Friedman, Jane E. "What's the Math Moral of the Story?" *Childhood Education* (Fall 1997): 33–35.

*Geometry and Geometric Thinking.* Focus Issue. *Teaching Children Mathematics* 5 (February 1999).

Griffiths, Rachel, and Margaret Clyne. *Books You Can Count On: Linking Mathematics and Literature.* Portsmouth, N.H.: Heinemann Educational Books, 1991.

Harris, Jacqueline. "Using Literature to Investigate Transformations." *Teaching Children Mathematics* 4 (May 1998): 510–13.

Hoban, Tana. *Shapes, Shapes, Shapes.* New York: Greenwillow Books, 1986.

Huellmantel, Pat. *Teach Mathematics with Children's Literature: Trade Books and Teacher Resource Books.* Flint, Mich.: Keith Distributors, 1999.

Hutchins, Pat. *Changes, Changes.* New York: Macmillan Publishing Co., 1987.

———. *The Doorbell Rang.* New York: William Morrow & Co., 1989.

Jacobs, Victoria R., Tom R. Bennett, and Cathy Bullock. "Selecting Books in Spanish to Teach Mathematics." *Teaching Children Mathematics* 6 (May 2000): 582–87.

Karp, Karen, E. Todd Brown, Linda Allen, and Candy Allen. *Feisty Females: Inspiring Girls to Think Mathematically.* Portsmouth, N.H.: Heinemann, 1998.

Lottridge, Celia Barker. *One Watermelon Seed.* Toronto: Oxford University Press, 1990.

MacDonald, Suse. *Sea Shapes.* New York: Harcourt Brace & Co., 1998.

McGrath, Barbara Barbieri. The *M & M's Brand Counting Book.* Watertown, Mass.: Charlesbridge Publishing, 1994.

Murphy, Stuart. *Elevator Magic.* New York: Harper Collins Publishers, 1997.

———. *Too Many Kangaroo Things to Do.* New York: Harper Collins Publishers, 1996.

National Association for the Education of Young Children (NAEYC). *NAEYC Position Statements.* Washington, D.C.: NAEYC, 1997.

National Board for Professional Teaching Standards (NBPTS). *The National Board for Professional Teaching Standards.* 2002. www.nbpts.org/about/coreprops.cfm (8 January 2003).

National Council of Teachers of English (NCTE). *Guidelines for the Preparation of Teachers of English Language Arts.* Urbana, Ill.: NCTE, 1996.

National Council of Teachers of Mathematics (NCTM). *Curriculum and Evaluation Standards for School Mathematics.* Reston, Va.: NCTM, 1989.

———. *Principles and Standards for School Mathematics.* Reston, Va.: NCTM, 2000.

Neuschwander, Cindy. *Sir Cumference and the First Round Table: A Math Adventure.* Watertown, Mass.: Charlesbridge Publishing, 1997.

Ohio Department of Education. *Model Competency-Based Mathematics Program.* Columbus, Ohio: State Board of Education and Ohio Department of Education, 1990.

Post, Thomas R., Arthur K. Ellis, Alan H. Humphreys, and L. JoAnne Buggey. *Interdisciplinary Approaches to Curriculum: Themes for Teaching.* Upper Saddle River, N.J.: Prentice Hall, 1997.

Ross, Elinor P. *Using Children's Literature across the Curriculum.* Bloomington, Ind.: Phi Delta Kappa Educational Foundation, 1994.

Russell, Dorothy S., and Elaine M. Bologna. "Teaching Geometry with Tangrams." In *Geometry for Grades K–6: Readings from the* Arithmetic Teacher, edited by Jane M. Hill, pp. 133–36. Reston, Va.: National Council of Teachers of Mathematics, 1987.

Smithsonian Astrophysical Observatory. "Guess My Shape: Resources for Teachers and Students." *Everyday Classroom Tools.* 1998. hea-www.harvard.edu/ECT/Guess/resources.html (7 January 2003).

Thiessen, Diane, and Margaret Matthias. "Selected Children's Books for Geometry." *Arithmetic Teacher* 37 (December 1989): 47–51.

Tompert, Ann. *Grandfather Tang's Story: A Tale Told with Tangrams.* New York: Crown Publishers, 1990.

Welchman-Tischler, Rosamond. *How to Use Children's Literature to Teach Mathematics.* Reston, Va.: National Council of Teachers of Mathematics, 1992.

Whitin, David. "Exploring Mathematics through Children's Literature." *School Library Journal* 38 (8) (August 1992): 24–28.

Whitin, David J., Heidi Mills, and Timothy O'Keefe. *Living and Learning Mathematics: Stories and Strategies for Supporting Mathematical Literacy.* Portsmouth, N.H.: Heinemann Educational Books, 1990.

Whitin, Phyllis, and David J. Whitin. *Math Is Language Too: Talking and Writing in the Mathematics Classroom.* Urbana, Ill.: National Council of Teachers of English, 2000.

# 15

# What Children's Literature Can Offer


## Jennifer M. Bay-Williams
## Sherri L. Martinie


> Let us read with method, and propose to ourselves an end to which our studies may point. The use of reading is to aid us in thinking.
>
> —*Sir Edward Gibbon*

CREATING OR finding a mathematics lesson that teaches important mathematics concepts and integrates ideas across mathematics content strands is an important yet challenging goal for teachers. Examples of rich contexts that facilitate integrating mathematics are described in the grade-band sections for the Connections Standard in *Principles and Standards for School Mathematics* (*Principles and Standards*) (NCTM 2000). These examples include filling a jar (prekindergarten through grade 2), running a snack shop (grades 3–5), making cranberry juice (grades 6–8), and positioning a guard dog in a triangular yard (grades 9–12) (NCTM 2000). These problem-solving contexts offer opportunities to integrate within and among the five content strands: Number, Algebra, Geometry, Measurement, and Data. Finding more explorations like these—ones that provide an interesting context for students and integrate mathematics ideas—can be difficult. However, literature, including fiction, nonfiction, and poetry, offers many rich contexts for integrated mathematical investigations. Many elementary level teachers are familiar with the growing collection of literature that can be used to launch mathematics lessons, but middle school teachers may not be aware of the potential that literature holds for teaching higher-level mathematics content. At all grade levels, teachers have a multitude of opportunities to integrate mathematics within the context of a good poem, story, or news article.

## INTEGRATING MATHEMATICS THROUGH LITERATURE

BEFORE WE could build a middle school lesson around literature, we first had to identify the mathematics. Paramount to the teacher's role of selecting mathematics investigations is developing a coherent curriculum that includes important mathematics content (NCTM 2000). Such a curriculum includes both assessing the mathematics learning needs of students and determining what content is important for them to learn. For middle school students, we knew that the study of rational numbers was a particularly difficult topic. The students often do not have sufficient opportunities to develop concepts related to rational numbers, particularly proportional reasoning. *Principles and Standards* (NCTM 2000, p. 217) notes the following:

> Facility with proportionality involves much more than setting two ratios equal and solving for a missing term. It involves recognizing quantities that are related proportionally and using numbers, tables, graphs, and equations to think about the quantities and their relationship. Proportionality is an important integrative thread that connects many of the mathematics topics studied in grades 6–8.

Proportionality spans the content strands. Ratios and proportions, linear equations, similarity, scaling, and probability all involve proportional reasoning.

Once we determined that proportionality was an important concept that our students needed more opportunities to explore, the problem became where to find rich contexts for developing proportional reasoning abilities. We began looking through children's literature. In the next section we describe the result—a sixth-grade mathematics lesson that uses a children's book to develop understanding of proportional reasoning as well as to integrate other important mathematical concepts. We also include a brief summary of other literature that supports developing proportional reasoning.

### How Big Is a Foot?

In the fictional story *How Big Is a Foot* (Myller 1962), the king wants to give the queen something for her birthday, but she already has everything.

He decides to give her a bed. However, no one in the kingdom has a bed because beds have not been invented yet. The story centers on building a bed of the correct size for the queen. The king uses his foot to measure the size of the bed that is needed and, through several other people, shares this information with the apprentice. The apprentice then uses his own foot, which happens to be smaller than the king's foot, to measure and build the bed. This measurement method results in a bed that is much too small for the queen. This story is commonly read to younger children as an introduction to using standard units of measure, but for middle school students, it offers an opportunity to explore proportional reasoning. In the sixth-grade lesson, the students studied the length, width, perimeter, and area of the queen's bed using both the king's foot and the apprentice's foot. They also used measurement as a way to explore proportionality.

After hearing the story read aloud, the students discussed the problem that the first bed was the wrong size. Next they considered whether the apprentice could accurately measure the bed using his own foot rather than using a measure of the king's foot. For example, what if the apprentice's foot was one-half the length of the king's foot? The students used a variety of approaches to explore the situation. Some students drew a row of king's feet next to a row of apprentice's feet for comparison. Other students created a chart showing the number of king's feet needed for a given length and the related number of apprentice's feet needed for the same length. Some students recognized that because the apprentice's foot was half the size of the king's foot, the apprentice needed to take twice as many steps. Exploring this situation laid the foundation for exploring a more complicated scenario—the apprentice's foot being two-thirds the size of the king's foot. Motivated and challenged, the students worked on this new measurement problem at length, incorporating several tools and strategies.

In one strategy, shown in figure 15.1, the students used thirds to solve the problem. The group explained that six of the king's steps contained eighteen thirds. They drew the apprentice's step to show that it used two of the eighteen each time, or a total of of nine steps. Several groups of students requested to use rectangular rods to implement a similar strategy. In these instances, they picked two rods, one two-thirds the length of the other (e.g., with Cuisenaire Rods, using the 6-rod for the apprentice's foot and the 9-rod for the king's foot).

Applying the relationship to the length of an actual foot was another strategy. One group created a life-sized model by lining up strips of tape that were the length of the king's foot—which happened to be twelve inches (i.e., the measure of a foot in standard units)—on the floor and

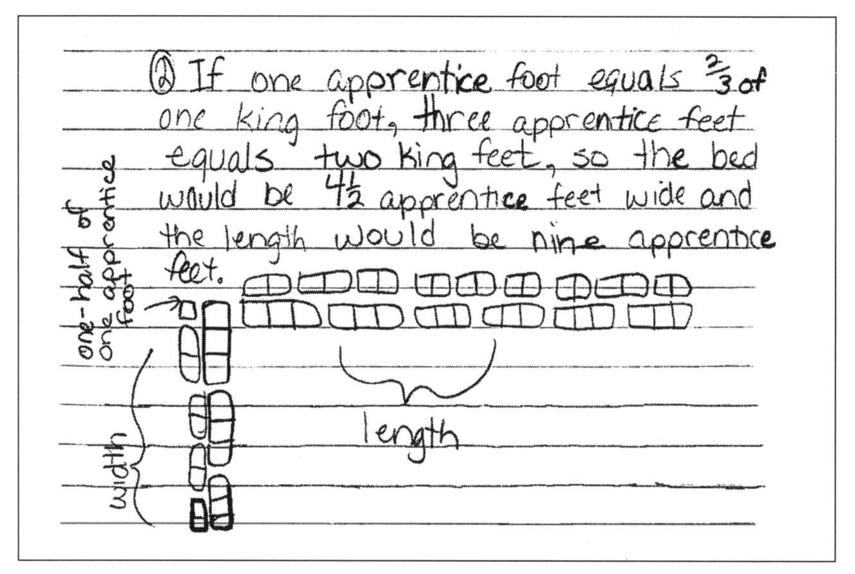


Fig. 15.1. Students' strategy using thirds

then measuring off strips of tape that were eight inches (two-thirds of a foot) long to represent the apprentice's foot. Then they placed the eight-inch strips next to the twelve-inch strips. The students saw that the apprentice needed to take nine steps to equal the king's six steps and four and one-half steps to equal the king's three steps. See figure 15.2.

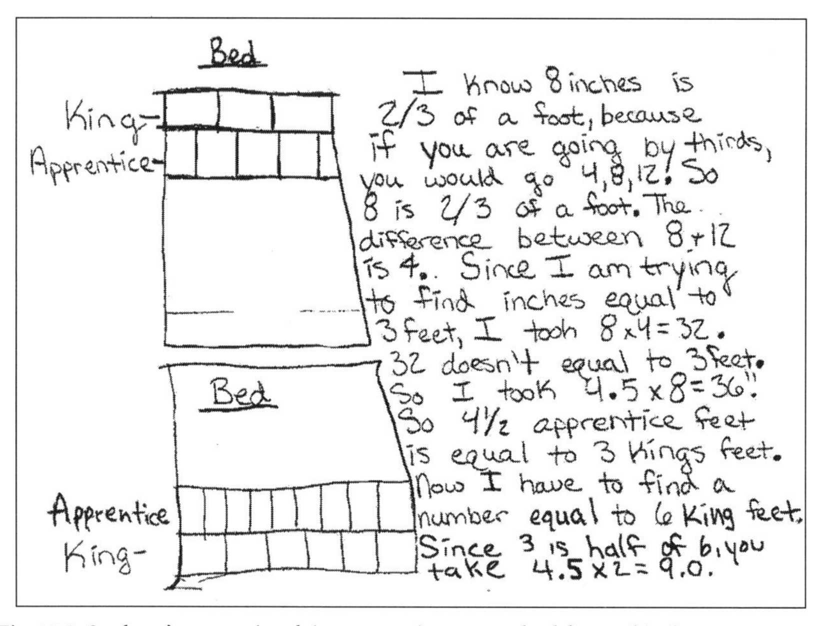


Fig. 15.2. Students' strategy involving conversion to standard feet and inches

In a third method, the students created a table listing the number of king's feet in one column and the number of apprentice's feet for the same length in the other column. The students used the data in the table to generalize the pattern (i.e., to determine the scale factor). In this strategy, the students were implicitly using the algorithm for the division of fractions. As each group shared its strategy for solving the problem, other groups could see alternative approaches, each of which involved proportional relationships.

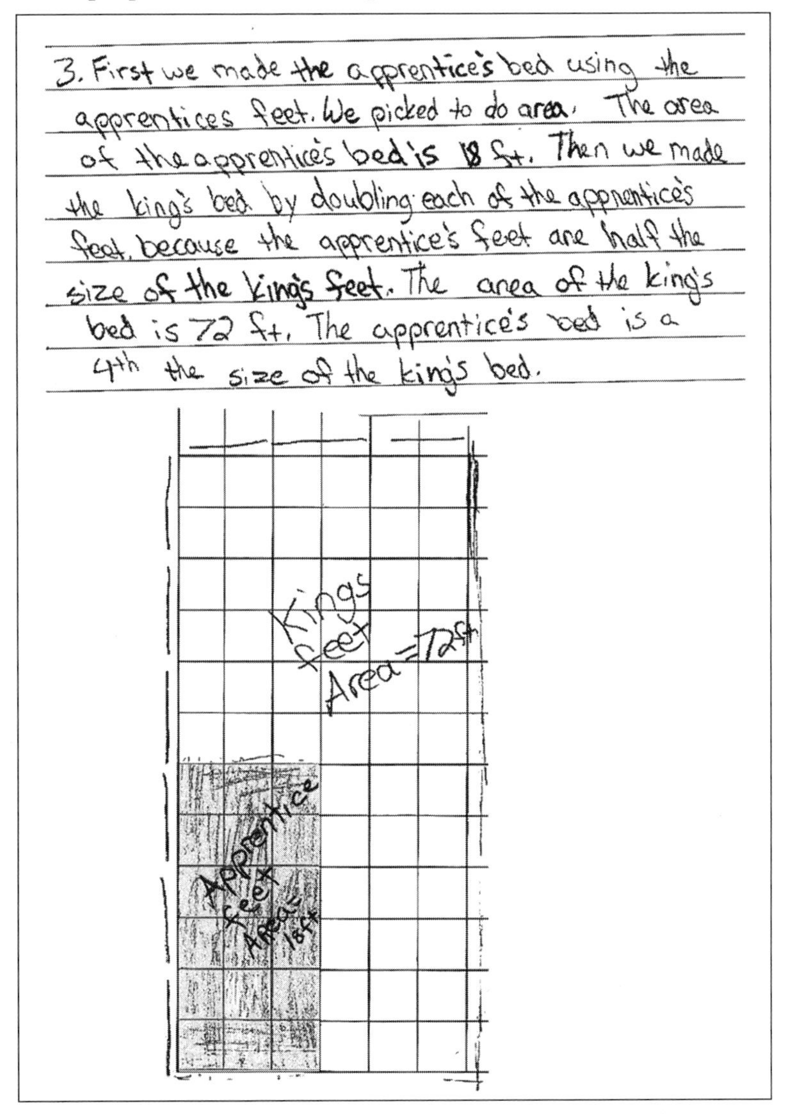


Figure 15.3. Student's exploration comparing area of the two beds when the apprentice's foot is one-half the size of the king's foot

As an extension, the students discussed how to measure the size of the bed other than using length and width. Suggested approaches included measuring the bed's diagonal, measuring its perimeter, and calculating the area of the top of the mattress. Each group selected one of these new measurements to explore. Figure 15.3 shows the work of a group that investigated the area using the original scenario of the apprentice's foot being half the size of the king's foot. These students were surprised to find that when the apprentice's foot is one-half the size of the king's foot, the area of the apprentice's bed is one-fourth the size of the king's bed. With this activity, the students were able to explore how the rate of growth for length is different from that for area.

Overall, in our efforts to have the students reason proportionally, literature offered a rich context. In this lesson, the students measured, worked with fractions, and developed tables in which they studied and generalized patterns. The open-ended problem-solving task embedded in the literature resulted in integration across the content strands.

## More Literature Links

Proportionality investigations can be done with many other books, particularly those that include comparisons. These books, for example, *How Big Is a Foot*, offer multiple opportunities to explore mathematics across the strands. In *Gulliver's Travels*, by Jonathan Swift (1947), students can compare a shirt sized for Gulliver with a shirt sized for the six-inch-tall people on Lilliput Island and can explore scale drawings and similarity, both important ideas in geometry. *Jim and the Beanstalk*, by Raymond Briggs (1970), involves fitting a giant with clothing and such items as eyeglasses. Shel Silverstein's poem "One Inch Tall" (1974) can also be used to explore scale factors, but this story involves scaling down, not up. In *Much Bigger Than Martin*, by Steven Kellogg (1977), Henry and his brother Martin post on the wall their measurements and the corresponding dates. Given the boys' height differences, the students can reason proportionally to determine the sizes of the boys' hands, feet, and arms. Measurement and patterns are central to these investigations. The book *If You Hopped Like a Frog*, by David Schwartz (1999), compares what a human could eat, see, and do if possessing the characteristics of various creatures (e.g., if you scurried like a spider, you could charge down an entire football field in just two seconds). Furthermore, in a lesson based on this story, the teacher could integrate data collection by having the students gather and graph information on the maximum speeds that different species can attain in comparison with their weights.

# Summary

With a mathematical goal in mind, teachers can select literature that fosters progress toward the goal as well as opportunities to integrate other mathematical ideas. Literature can enrich the student's "pool of mathematical ideas" that can be applied to solving a given task (Whitin and Whitin 2000, p. xii). In the lesson described in this chapter, the many solution strategies used to determine the dimensions for building the queen's bed are evidence of such enrichment. Literature also offers a "human perspective about mathematics" (Whitin and Wilde 1995). Problems that arise out of a story can build a solid contextual foundation for learning, understanding, and using functional mathematics. Students' interest in the story also facilitates their problem posing. This effect was illustrated in the lesson above when the students selected various measurements to investigate the size of the beds. Furthermore, literature prompts students to use the context to develop problem-solving methods that make sense to them—as they did in the class described—rather than to struggle with poorly understood procedures.

The literature discussed here is only a small sample of the numerous books and resources that can be used to explore connections within mathematics and with the world. The first step in finding appropriate literature is to define the mathematical goals, then look for supporting material. For us, once we determined our goal to be developing proportionality, we scanned books and bibliographies for books that could support our goal. One helpful resource is *Wonderful World of Mathematics* (Thiessen, Matthias, and Smith 1998), an annotated bibliography of literature that can be used to teach mathematics. It describes the literature in enough detail for readers to determine how each story might fit with their mathematical goals. Other resources include *Read Any Good Math Lately?* (Whitin and Wilde 1992), *It's the Story That Counts* (Whitin and Wilde 1995), *Books You Can Count On: Linking Mathematics and Literature* (Griffiths and Clyne 1988), and the *Math and Literature* series (Burns 1993; Sheffield 1994; and Bresser 1995). As you select the mathematics curriculum that is appropriate for your students, consider using literature to integrate mathematical ideas and to help your students make sense of the mathematics.

## References

Bresser, Rusty. *Math and Literature (Grades 4–6)*. White Plains, N.Y.: Math Solutions Publications, 1995.

Briggs, Raymond. *Jim and the Beanstalk.* New York: Putnam & Grosset Group, 1970.

Burns, Marilyn. *Math and Literature (K–3): Book One.* White Plains, N.Y.: Math Solutions Publications, 1993.

Griffiths, Rachel, and Margaret Clyne. *Books You Can Count On: Linking Mathematics and Literature.* Portsmouth, N.H.: Heinemann, 1988.

Kellogg, Steven. *Much Bigger Than Martin.* New York: Dial Books for Young Readers, 1976.

Myller, Rolf. *How Big Is a Foot?* New York: Bantam Doubleday Dell Publishing Group, 1962.

National Council of Teachers of Mathematics (NCTM). *Principles and Standards for School Mathematics.* Reston, Va.: NCTM, 2000.

Schwartz, David M. *If You Hopped Like a Frog.* New York: Scholastic Press, 1999.

Sheffield, Stephanie. *Math and Literature (K–3): Book Two.* White Plains, N.Y.: Math Solutions Publications, 1994.

Silverstein, Shel. *Where the Sidewalk Ends.* New York: Harper Collins Publishers, 1974.

Swift, Jonathan. *Gulliver's Travels.* 1726. Reprint, adapted by James Riordan, New York: Oxford University Press, 1998.

Thiessen, Diane, Margaret Matthias, and Jacquelin Smith. *The Wonderful World of Mathematics: A Critically Annotated List of Children's Books in Mathematics.* Reston, Va.: National Council of Teachers of Mathematics, 1998.

Whitin, David J., and Sandra Wilde. *It's the Story That Counts: More Children's Books for Mathematical Learning, K–6.* Portsmouth, N.H.: Heinemann Educational Books, 1995.

———. *Read Any Good Math Lately?* Portsmouth, N.H.: Heinemann Educational Books, 1992.

Whitin, Phyllis, and David J. Whitin. *Math Is Language Too: Talking and Writing in the Mathematics Classroom.* Urbana, Ill.: National Council of Teachers of English, 2000.

16

# Rocket Racers and Submarines: Two Australian Case Studies of Curriculum Integration in Practice

Fiona Budgen
John Wallace
Léonie Rennie
John Malone

MOST AUSTRALIAN children enter high school around age 12. This transition heralds a number of changes for students—most noticeably, they move from the more intimate environment of an elementary school to a situation where they are taught highly discipline-based content by many teachers in many classrooms. These changes can make the early years of high school potentially stressful and alienating and, as a consequence, are of considerable educational concern (Cormack 1996; Speering and Rennie 1996). Teachers in many Australian schools are faced with the complex problem of how to motivate and engage these early adolescent students in meaningful learning.

Partly—but certainly not exclusively—in response to these concerns, teachers and administrators in Australia have become increasingly interested in curriculum integration, particularly the opportunity to incorporate mathematics across the curriculum. Past curricular practices in mathematics have disadvantaged and alienated many students. Integration appears supported by research that suggests a need to move away from rote-learning procedures and to adopt curricular models through which students understand the processes and appreciate the relevance of mathematics. Alper et al. (1996, p. 20) explain that when "students construct ideas in context, instead of just memorizing definitions, mathematical concepts and methods have real meaning to them." Several

Australian state curriculum documents have been developed that advocate an integrated approach to teaching and learning (e.g., the *Curriculum Framework* from the Curriculum Council of Western Australia [1998] and the *Guidance Statement on Curriculum Integration* from the New South Wales Board of Studies [1996]), and schools around Australia are being organized to facilitate integrated and interdisciplinary approaches to curriculum delivery.

## THE CASES FOR AND AGAINST INTEGRATION

AT THE heart of integration is the notion that integrated curricula can motivate students by promoting wholeness and unity rather than separation and fragmentation. Beane (1991) argues that traditional subject disciplines are "territorial spaces carved out by academic scholars for their own purposes" (p. 9) and that their boundaries limit students' access to broader meanings. An integrated curriculum, its advocates propose, leads to "education for insight," resulting in learning rich with connection-making and the flexible use of knowledge (Perkins 1991). Several advantages for both teachers and students are outlined by Clark and Clark (1994). They suggest that teachers experience greater satisfaction, renewed energy, and excitement when working together on integrated curricula. Furthermore, they say that students become more involved and enthusiastic and demonstrate less competition and more cooperation. However, the same authors point out the pitfalls of, and barriers to, integration, including the following:

- Concerns that essential subject matter will be compromised
- The perceived lack of teachers' expertise and training in using an integrated approach
- A lack of acceptance and support in the schools and community for an integrated curriculum
- The increased amounts of time that teachers need to collaborate in developing an integrated curriculum

The arguments—and counterarguments—for integrating science, mathematics, and technology are summarized in a special issue of *School Science and Mathematics* (Lederman and Niess 1998). Proponents of an integrated curriculum argue that disciplines are historical creations that fragment and compartmentalize knowledge, constraining teachers' ability to carry out learning activities that reflect or connect with the way in which children perceive the world (Hargreaves, Earl, and Ryan 1996).

Conversely, advocates of a traditional, discipline-based approach argue that robust understandings of important phenomena and concepts depend on the study of disciplines and the methods and approaches of those disciplines (Hatch 1998). Isaacs, Wagreich, and Gartzman (1997) acknowledge that the basic concepts of mathematics and science can be obscured in an integrated context, and George (1996) argues that certain topics will be interwoven into an integrated curriculum far more frequently than others. Such debates generate important questions, including the following:

- Can teachers create a classroom environment that has the best of both worlds—promoting learning both within and across disciplines? (Wallace et al. 2001)
- Does integration increase the likelihood that learning opportunities will become learning outcomes?
- What is the influence of integrated teaching on students' motivation?
- Do students flexibly apply their knowledge when and where it is needed?

This chapter explores some of these questions by describing two technology-based projects that integrate mathematics, science, and technology.

## THE RESEARCH

THE FOLLOWING two case studies produced different levels of success in the classroom and illustrate both advantages of, and pitfalls in, this kind of curricular approach. The intensive study, conducted at a single school site, involved one group of eighth-grade students who undertook two technology-based projects under the direction of the same two teachers. One teacher was responsible for the mathematics and science, whereas the other taught technology—a curriculum area in which students apply knowledge, skills, and resources to solve technological problems in ways that meet the needs of individuals, societies, and environments. Each case study was conducted over a full school term with the opportunity for "prolonged engagement" and "persistent observation" (Guba and Lincoln 1998), allowing researchers to describe in detail the structural, pedagogical, and learning characteristics of each setting. Several complementary monitoring tools were employed, including direct observation of performance tasks, note taking in the field, interviews, concept mapping, audiotaping, photography, review of learning journals, and analysis of such artifacts as students' portfolios and teachers' notes. The focus was on actual classroom tasks and providing sufficient data to ensure a "rich description" (Denzin and Lincoln 2000) of the events that unfolded in each case study.

## Case Study 1: Rocket Racers

This project took place during the second school term and involved students in designing and building model rocket cars powered by compressed air.

## How the Project Was Approached

The technology teacher instructed students to design and build a vehicle powered by compressed air. The goal was to produce a vehicle capable of travelling the fastest along a guide-track 33 meters in length. A sample of the track material (made from 25 millimeter-wide "U-shaped" plastic channel) was displayed so that the students could take measurements and consider suitable guidance systems. The teacher showed the students the construction materials and tools at their disposal. The only item that the students were asked to supply was a clean, empty 600 milliliter plastic soft-drink bottle.

For this project, the students were expected to follow a four-step process: (1) design, (2) make, (3) evaluate, and (4) change. After some discussion about the design, the class established a few factors that would require consideration. These included the following:

- Aerodynamics—what shapes can lessen wind resistance?
- Materials—how can the vehicle be made lightweight without compromising its strength?
- The track—what do you need to know about the track before you build your vehicle?
- Guidance system—how will the vehicle follow the track?

Meanwhile, the science and mathematics teacher introduced the science topics related to energy transfer, including force, work, power, and efficiency. The students' tasks included naming and describing three forces acting on the rocket racer and drawing a diagram to show the direction in which these forces were acting. The students also had to explain why all the potential energy from the compressed air was not transferred to the kinetic energy of the rocket racer and describe how factors affecting friction would influence the design of the vehicle.

The process of building the vehicle required students to use mathematics in measuring heights, widths, lengths, and diameters. The teacher reviewed the definitions of radius, diameter, and circumference as well as the process of drawing the net of a cone and the relationship between mass and volume of water. She also reviewed the concept of scale because all designs had to be drawn on a scale of 1:2.

The main mathematical learning involved planning for the collection of data during the testing phase. The class determined three variables to consider during the trials: the mass of the vehicle, the volume of the water added, and the diameter of the hole in the bottle's screw cap. Each group was asked to design a table in which they could collect and record the data required for analysis to improve the performance of their vehicles. During the testing process, groups used stopwatches to time their vehicles. If the vehicle did not travel to the end of the track, the actual distance traveled was measured using a trundle wheel, and the time taken for the vehicle to travel that distance was recorded. These measurements were later used to calculate the speeds attained by students' vehicles and to produce computer-generated graphs of the results.

## Designing and Building the Racers

The students were eager to begin building their vehicles but reluctant to spend time on the design phase of the process. Their designs showed that they had given little consideration to the points discussed earlier—how to lessen wind resistance, how to make the vehicle lightweight without compromising its strength, and how to make it stay on the track. The designs were very simple and remarkably similar (see fig. 16.1). Most of the designs consisted of a chassis of rectangular aluminium bent to form shallow sides through which the students drilled holes to accommodate the axles. The soft-drink bottles were taped to the chassis.

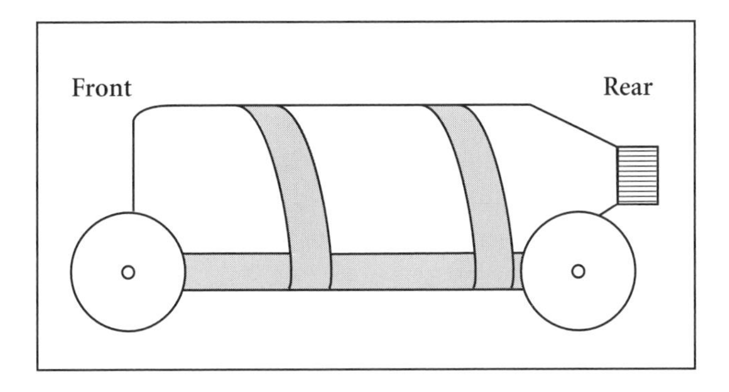


Fig. 16.1. Early design for a rocket-car racer

## Testing the Rocket Racers

The testing of the first prototypes revealed many design problems. Most of the vehicles traveled only a short distance before stopping or leaving the track. The students began to realize that this project was more complex than they had first imagined. Some of the problems that they encountered included the following:

- Instances of vehicles rubbing on the track or becoming suspended on it—this problem occurred because students had not checked the height measurements. Furthermore, they had either chosen wheels that were too small or had not considered the consequences of drilling the holes on the chassis for the axles.
- Instances of chassis and wheels contacting each other—the teacher pointed out to most groups that this contact was causing friction and that students needed to insert a spacer on the axles between the wheels and chassis.
- Instances of the guidance system being inadequate for the procedure—some groups tried setting the wheels on one side of the vehicle in the channel of the guide track (see fig. 16.2), but this position did not produce good results. Others attached a system that created too much friction or found that their system was too fragile to withstand the testing procedure.

Fig. 16.2. Rocket racer with wheels set in guide track

The plan had been for the science and mathematics teacher to be present during the vehicle trials, but on each occasion she was called out to cover for teachers who were absent. This situation left the technology teacher to deal single-handedly with the emerging problems of vehicle design and performance. After the second day of trials, students appeared to have exhausted their ideas about how to improve their vehicles. None of the groups produced a vehicle that would reach the end of the track or perform consistently.

At the beginning of the third day of trials, the technology teacher built a vehicle of his own. The students instantly started asking such questions as "Is that the winning car? Is that the right car?" They appeared to believe that only one correct design solution existed and, before they saw the vehicle being tested, began to modify their own vehicles to match the teacher's design. They frantically drilled holes in the chassis to make it lighter. Four groups attached plastic nose cones to their cars, but only two could explain that the purpose was to improve the aerodynamics of the vehicle. Everyone came to watch the teacher's vehicle being tested, but the car traveled only a short distance along the track before stopping. The teacher measured the distance, which was about five meters, then picked up his car and told the students, "I think it might be too heavy. I added 300 milliliters of water, so I'll try it with 200 milliliters next time and see what happens." He continued to systematically change a variable, and then test the car, each time recording the results on the board. This approach produced the desired effect on the students. They stopped making random changes to their own vehicles and began to work methodically. By the end of the lesson, all the vehicles had been improved and several were consistently reaching the end of the track.

## ASSESSMENT

THE STUDENTS were assessed on the quality of their record-keeping as well as on the finished vehicle's performance. On completion of the project, the teacher gave each student a photocopied list of essential terms to be linked in the concept map and encouraged the students to add any other necessary terms to make the concept map complete. The teacher explained that the purpose of the exercise was to demonstrate how much students had learned from working on the rocket-racer project. The groups cut out the words on the list and discussed, arranged, and rearranged the layout of the words on the page until they were satisfied that the best layout had been achieved. Then they pasted the words onto the page and wrote in the linking constructs.

The concept maps were used to enhance teachers' judgments about the quality and depth of learning that had taken place. The students who produced the winning car also produced a very well organized and well linked concept map. One student who had worked individually produced a car that performed reasonably well to begin with but, as she made changes and the vehicle's performance deteriorated, seemed unable to modify the car for the better. Her concept map revealed that she had a very limited understanding of what she was doing, and it seems likely that her early successes were attributable more to luck than design. One group added an animal-shaped superstructure to their car; this tactic did not enhance the car's aerodynamic qualities and added considerably to its mass, yet the group produced the strongest concept map and showed a sound understanding of the relationships among the given terms (see fig. 16.3). Those students clearly understood the concepts involved in the task, but that level of understanding had not been apparent when they were experimenting with their vehicle.

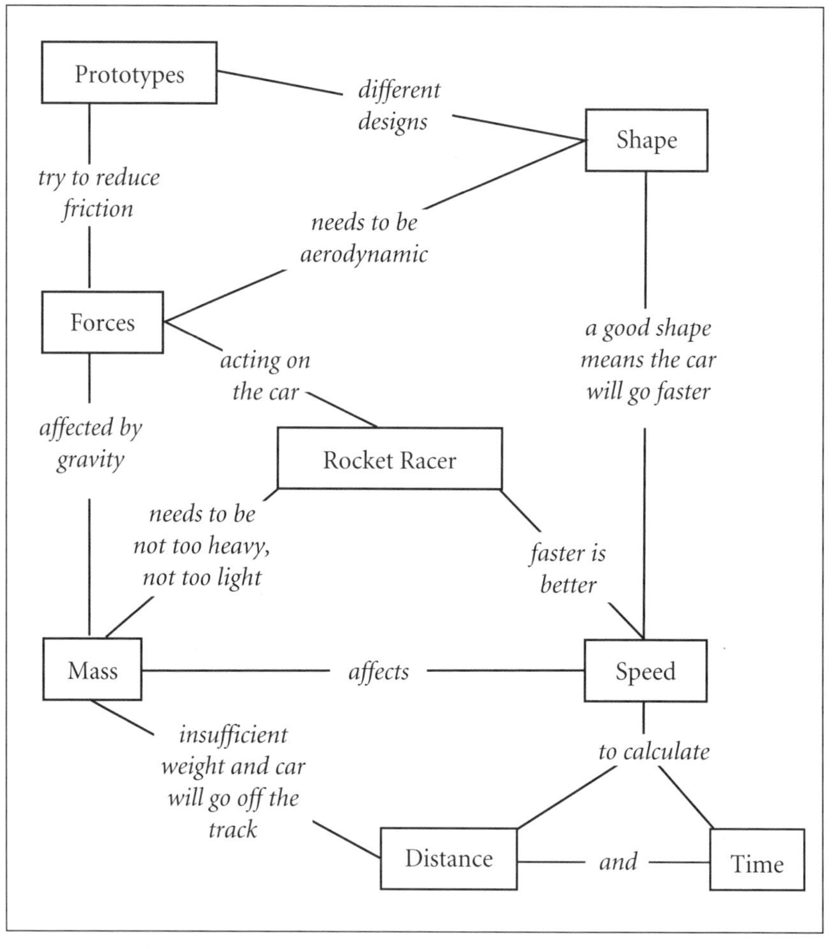


Fig. 16.3. Sample concept map

# DISCUSSION

THE DISCUSSION of the rocket-racer case study revolves around the questions that this chapter posed previously about learning outcomes, students' motivation, and application of knowledge.

- *Learning outcomes.* Given the time and effort put into this project by the teachers, the level of learning revealed by the concept-mapping activity was not encouraging. Although some evidence of learning was exhibited and a few of the students showed a robust understanding of the concepts involved, the outcomes for many of the students were disappointing. Of the eight groups and one individual who completed concept maps, two groups failed to note a connection between shape and speed, and three groups failed to note a connection between mass and speed. Only two groups noted the effects of friction anywhere in their concept maps, and three groups were unable to link any of the forces acting on the vehicle. None of the groups added any terms to the concept map other than those that had been provided by the teacher.
- *Motivation.* The lack of learning was especially apparent for the two or three behaviorally difficult students in the class. These students described the task as boring. Their levels of motivation remained low throughout the project, and they achieved little. In fact, the levels of motivation demonstrated by all the students were generally quite disappointing given the nature of the project, the enthusiasm and patience of the teachers, the materials and equipment at students' disposal, and the opportunity that the project presented to work on a collaborative, practical activity. The students seemed unmotivated from the beginning; for instance, although the students were asked to supply the empty soft-drink bottles for their vehicles, no bottles were supplied.
- *Application of knowledge.* The students seemed to have difficulty making connections between what they were learning in science and mathematics and the problems they had to solve in technology. They seemed unwilling or unable to recognize the usefulness of their knowledge or apply it. For example, although investigating variables was a major component of the mathematics teaching, not until the third day of testing—when the technology teacher modeled the process— did the students apply this skill. The students measured the dimensions of the track for the vehicle but failed to recognize the implications of these measurements for the vehicle's design. None of

the students attempted to give their vehicles a more aerodynamic shape, and many failed to consider the effect of friction in spite of this topic's explicit coverage during the science lessons. The claimed advantages of integration—that it results in rich, connection-making learning and in students who are more cooperative, involved, and enthusiastic—were not evident during this project.

## Case Study 2: Submarines

This project took place later in the year, during the fourth school term, and involved the same group of students—but this time, in building and testing model submarines that had been devised and designed by the technology teacher. Figure 16.4 shows a completed model. Unlike in the previous project, the students were not required to design their vessels; however, a number of students still initially reacted to the project with resistance. The technology teacher strove to hold the students' attention during the introductory lesson.

Fig. 16.4. Model submarine

## How the Project Was Approached

The technology teacher showed the students the submarine design and provided a cutaway model so that students could see all the operating parts and how they worked (see fig. 16.5). The submarine model allowed water to enter through the lower valve and air to escape through the top valve, which was held open by a wooden float. As the water level rose, the submarine sank. The float holding open the top valve rose with the water, eventually allowing the valve to close. When water entered the dry-ice canister, the dry ice started to vaporize more quickly, filling the main chamber with carbon dioxide gas and forcing the water out through the copper pipe, thus propelling the submarine forward. The submarine's buoyancy increased because the water had been displaced by carbon dioxide, and it returned to the surface for the cycle to repeat.

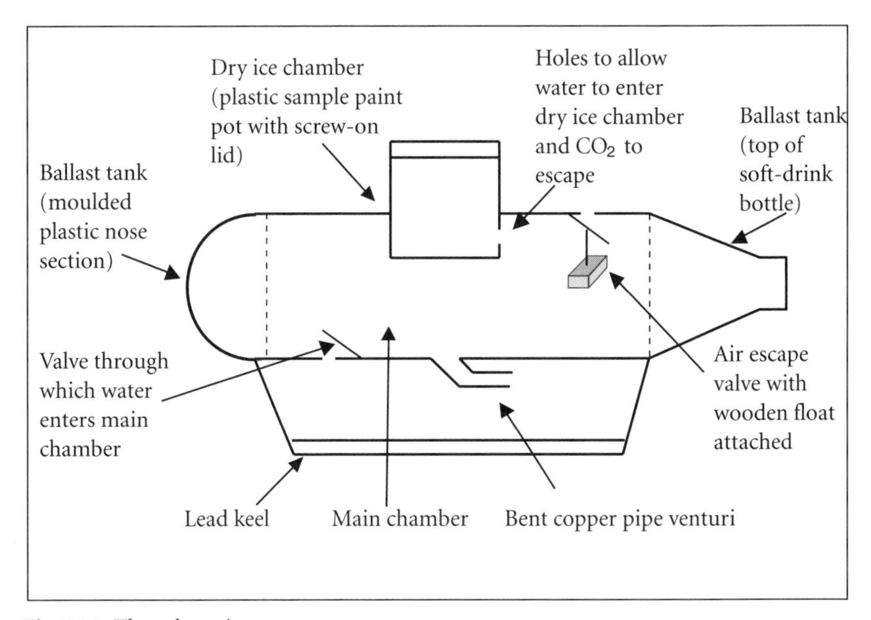


Fig. 16.5. The submarine

Under the supervision of the technology teacher, the students built their own submarines. Meanwhile in science, they covered such topics as floating, sinking, buoyancy, Archimedes' principle, density, states of matter, air pressure, and jet propulsion. The students also researched the history of submarines and the parts of a submarine. The project included two major mathematics components. The first required students to examine measuring units and instruments and to investigate the amount of measurement error that is tolerable in various circumstances. The importance of accuracy in measurement quickly became apparent during the building process because components that did not fit well together created gaps that were difficult and messy to seal. The second mathematics component required students to read from scale drawings, a skill that would be required to follow the provided instruction sheets for each of the submarine's components—the keel, the conning tower, the valves, and the fore and aft ballast tanks. The students were also asked to determine how to rule two lines equidistant from each other along the upper and lower surface of the submarine's cylindrical hull.

As a result of previous experiences in running this project, the science and mathematics teacher also dedicated an entire lesson to trouble-shooting: "What can you do if ____ happens?" Past experiences had taught the teaching staff that, in spite of doing a number of projects in which students were asked to (1) make, (2) evaluate, and (3) change, many of them gave up very quickly if they did not have instant success.

## Testing the Vessels

The students conducted their tests at a nearby swimming pool. Before starting, the students were reminded of the variables (ballasts in fore and aft tanks and mass of lead on the keel) that they would have to measure and trim to get the submarine to work correctly. These measurements had to be taken and recorded before the submarines entered the water. The first tests were done without dry ice and were conducted to ensure that the submarine would sink nose-first and sit level when it settled on the bottom of the pool (see fig. 16.6). Once the students were sure that their submarines were correctly trimmed, they proceeded with the test using dry ice. Most groups had to solve a few problems at the outset, but once the submarines were operating successfully, they glided through the water, diving and resurfacing about four or five times before having to be refueled.

Fig. 16.6. Dive! Students tested their submarines at a local swimming pool.

# ASSESSMENT

PART OF the assessment was based on the finished product. However, the students also completed an activity booklet that included all the classroom activities relating to the mathematics and science, as well as such independent research activities as investigating why many fish have swim bladders and how they work. The students also completed a technology report, which allowed the teacher to evaluate how accurately each group had set the variables, observed what had happened, recorded the results, and made necessary changes. Finally, during the building process, the students kept a learning journal in which questions were posed, the students gave answers, and then the teachers responded. This journal proved to be an effective way of keeping track of each student's level of understanding about the project and addressing any misconceptions that arose along the way.

# DISCUSSION

- *Learning outcomes.* The learning journals provided evidence that learning was taking place and were also a useful means of facilitating learning. The questions probed the students' understandings, and the teachers were able to respond quickly to any emerging misconceptions or to help students who had not grasped a concept. One commonly occurring misconception was that the submarine would be propelled by air escaping through the venturi (a constriction in the pipe to speed up water flow). For a number of students, this misconception was addressed through learning-journal dialogues that generally followed a pattern such as this one:

  *Teacher:* What would you expect to see if air was coming out from the venturi?

  *Student:* Bubbles.

  *Teacher:* The space above the venturi is filled with water, so how does the air reach the venturi?

  *Student:* Is it water that comes out?

  *Teacher:* Look closely when you test your submarine. If you see bubbles, then there must be air; but if there are no bubbles, then it must be water.

  Incomplete answers could be expanded through these written dialogues, but occasionally, a student was unable to offer any answer to the question posed. In these circumstances, the teacher was able to give the student some individual attention. Students' sound understanding of the process by which the submarine operated was fundamental to their being able to solve problems on the vessel test, so it was vital to ensure that the learning had taken place before the students took their submarines to the pool.

- *Motivation.* When the project was first introduced to the class, the students appeared to show little motivation or interest. However, as the building process began, their level of involvement increased, and by testing day, every group had managed to produce a vessel ready to be launched.
- *Application of knowledge.* The students would have been unlikely to remain focused on their tasks or to persevere to make their submarines work without the lesson devoted to troubleshooting. When problems were encountered, the teachers were able to guide students to figure out solutions for themselves by asking such questions as "Why will your submarine not resurface if you are seeing bubbles escaping from the top valve?" "How can you get that valve to seal?"

## CONCLUSION

WHY DID these two projects produce such different outcomes in terms of students' learning and motivation? First, the two projects were organized differently—the rocket racer involved a (1) design, (2) make, (3) evaluate, and (4) change process, whereas the submarine project did not require any design input from the students. As the rocket-racer project showed, designing was certainly an area of weakness for these students—a fact that they themselves acknowledged in post-project interviews. When asked whether they would do anything differently, the majority of the students commented that they would give more attention to the design and planning stage:

*Matt*: We didn't know until we put it together that we needed more planning.

*Josh*: We should have thought about it more. We were a bit ahead of ourselves.

*Louise*: Think of where things are going to be, not just parts of it.

The students might have preferred to avoid the design phase, but the design phase was unlikely to have been the only factor influencing the success of the projects. Second, the timing of the project could have affected the outcomes. Because the second project was carried out at the end of the year, it could be argued that the students had become more familiar with the style of integrated learning and were more responsive to it. However, the submarine project initially met with disinterest from the students. Furthermore, conducting a project in the final term of a school year (the Christmas term in Australia) is fraught with setbacks. Several lessons were cancelled or rescheduled because of competing demands on the timetable and, with so many other activities happening, the students could easily have lost interest in the project.

Another difference between these two projects was in the strength of the connections among the science, mathematics, and technology concepts. During the rocket-racer project, the science and mathematics teacher was unexpectedly not able to play a role in the testing phase, so the integration was left entirely to the technology teacher. This situation may have obscured the interconnectedness of the subjects or even unintentionally downplayed the application of science and mathematics—making it less likely that any transfer of learning would take place. During the submarine project, however, the science and mathematics teacher was much more involved in the preparation of the submarines and the testing process.

The two case studies show that the success of integrated programs cannot be assumed. Both projects were conceptually and pedagogically sound. They were well supported by materials and by teachers who were committed to the notion of integration. The school administration supported teachers' collaboration and the scheduling of mathematics, science, and technology as an amalgamated unit. However, such factors as students' willingness and readiness to learn, their maturity, and their perceptions about the interconnectedness of subjects gained through witnessing teacher collaboration may have contributed to the outcomes.

We conclude that the success of integrated teaching must be attributed to a complex, and sometimes subtle, web of factors, including high-quality materials; a supportive culture; and qualified, committed, and collaborative teachers. It is possible that students who have trouble recognizing the connections among integrated subjects are helped by seeing the teachers jointly conduct some of the lessons; therefore, factors such as students' receptiveness, prior experience, and perceptions must also be taken into account.

## REFERENCES

Alper, Lynne, Dan Fendel, Sherry Fraser, and Diane Resek. "Problem-Based Mathematics—Not Just for the College-Bound." *Educational Leadership* 53 (May 1996): 18–21.

Beane, James A. "The Middle School: The Natural Home of Integrated Curriculum." *Educational Leadership* 49 (October 1991): 9–13.

Clark, Donald C., and Sally N. Clark. "Meeting the Needs of Young Adolescents." *Schools in the Middle* 4 (September 1994): 4–7.

Cormack, Phil. *From Alienation to Engagement: Opportunities for Reform in the Middle Years of Schooling.* Canberra: Australian Curriculum Studies Association, 1996.

Curriculum Council of Western Australia. *Curriculum Framework.* Perth, Australia: Curriculum Council of Western Australia, 1998.

Denzin, Norman K., and Yvonna S. Lincoln. "The Discipline and Practice of Qualitative Research." In *Handbook of Qualitative Research,* 2nd ed., edited by Norman K. Denzin and Yvonna S. Lincoln, pp.1–28. Thousand Oaks, Calif.: Sage Publications, 2000.

George, Paul. "Arguing Integrated Curriculum." *Education Digest* 62 (November 1996): 16–21.

Guba, Egon G., and Yvonna S. Lincoln. *Fourth Generation Evaluation.* Newbury Park, Calif.: Sage Publications, 1998.

Hargreaves, Andy, Lorna Earl, and Jim Ryan. *Schooling for Change: Reinventing Education for Early Adolescents.* London: Falmer Press, 1996.

Hatch, Thomas. "The Differences in Theory That Matter in the Practice of School Improvement." *American Educational Research Journal* 35 (spring 1998): 3–31.

Isaacs, Andrew, Phillip Wagreich, and Martin Gartzman. "The Quest for Integration: School Mathematics and Science." *American Journal of Education* 106 (November 1997): 179–206.

Lederman, Norman G., and Margaret L. Niess. "5 Apples + 4 Oranges = ?" *School Science and Mathematics* 98 (October 1998): 281–84.

New South Wales Board of Studies. *Guiding Statement on Curriculum Integration.* Sydney: New South Wales Board of Studies, 1996.

Perkins, David N. "Educating for Insight." *Educational Leadership* 42 (October 1991): 4–8.

Speering, Wendy, and Léonie Rennie. "Students' Perceptions about Science: The Impact of Transition from Primary to Secondary School." *Research in Science Education* 26, no. 3 (1996): 283–98.

Wallace, John, Léonie Rennie, John Malone, and Grady Venville. "What We Know and What We Need to Know about Curriculum Integration in Science, Mathematics, and Technology." *Curriculum Perspectives* 21 (April 2001): 9–15.

# Part 4

## Challenges, Roles and Responsibilities, and Implications

17

# Meeting State Standards with Integrated Problem-Solving Curricula

P. Mark Taylor and James E. Tarr

STANDARDIZED TESTING in the United States can be traced as far back as 1845, at which time the Boston School Committee decided to test the city's students to prove that the schools were doing a good job with their state funding. However, both the Boston School Committee and Horace Mann, the secretary of the State Board of Education in Massachusetts, were deeply disappointed in the results (Kilpatrick 1992).

Most teachers do not remember a time when state testing programs were not major factors in determining curricula and selecting textbooks. Furthermore, in many areas, state curriculum frameworks largely determine local school district curricula and schools seek textbooks that best reflect their state's achievement test. Publishing companies wisely maximize their profits by producing textbooks designed to satisfy standards and assessments for as many states as possible. As a result, these textbooks ultimately include more topics than mathematics teachers could possibly cover in any given school year. In this process, state curriculum frameworks and testing programs have led unintentionally to textbooks that are unfocused and unmanageable. Such curricular programs are the foundation on which the U.S. mathematics curriculum was built. Hence, the recent condemnation of the U.S. mathematics curriculum as "a mile wide and an inch deep" and "covering all areas but emphasizing none" by the national research coordinator of the Third International Mathematics and Science Study was not surprising (Silver 1998).

## BREAKING THE CYCLE WITH AN INTEGRATED CURRICULUM

HOW, THEN, are mathematics teachers to deal with such an extensive curriculum? Many address the problem by teaching as fast as they can without losing students along the way. Others attempt to arduously select which topics to teach and which skills— regardless of their value—to emphasize and practice in the weeks prior to the state assessment. The cycle that has developed (see fig. 17.1) starts with multiple state standards and accountability measures and leads to a broad, unfocused curriculum that can be seen in many current textbooks. When school districts and teachers use these wide-ranging textbooks to try to meet state standards, the practice often results in "teaching to the test" and students' gaining only superficial knowledge of an incoherent mathematics curriculum. Ironically, teaching to the test often results in poor test scores. States, in turn, react to their students' poor test scores by adjusting their standards or accountability system.

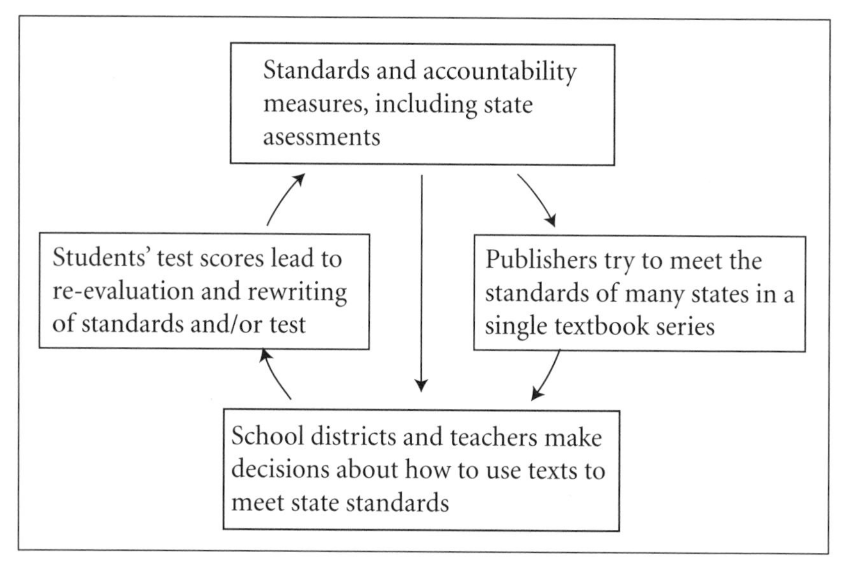


Fig. 17.1. The conventional cycle

This long-established cycle was broken only recently when some states fundamentally changed their assessment practices. In particular, some states began to emphasize problem solving, reasoning, and communication in their mathematics achievement tests. Moreover, some states changed their test formats to include open-ended and free-response items in addition to standard multiple-choice questions. A major factor in this change was the introduction of the *Standards* trilogy by the National Council of Teachers of Mathematics (NCTM). In 1989, *Curriculum and Evaluation Standards for School Mathematics* (NCTM) offered a new vision for school mathematics and articulated shifts in what mathematics content should be emphasized. Subsequent NCTM publications, *Professional Standards for Teaching Mathematics* (1991) and *Assessment Standards for School Mathematics* (1995), suggested changes in mathematics teaching and student assessment. Collectively, these three documents set the stage for sweeping changes in teachers' ideas about what mathematics should be taught, how it should be taught, and how it should be assessed (Lappan 1999).

Following the release of *Curriculum and Evaluation Standards*, publishers of mathematics textbooks rushed to claim that their products embodied tenets of the *Standards.* Few publishers, however, made significant attempts to streamline the curriculum by integrating mathematics topics the way that NCTM recommended. Without significant interventions, the NCTM *Standards* would have entered the conventional cycle (see fig. 17.1) and had only a minor effect on school mathematics.

In 1992, however, the National Science Foundation (NSF) called for curriculum directors—independent of textbook publishers—to develop mathematics curricula that would embody tenets of the NCTM *Standards.* To accomplish this ambitious goal, NSF funded three elementary school curriculum projects, five middle school projects, and five high school projects. The developers of each curriculum project worked independently because the intent was not to establish a federal mathematics curriculum but rather to allow individual project developers the liberty to build a curriculum that reflected their vision of the NCTM *Standards.* Instead of focusing on any one set of state frameworks, these curricula focused on the coherent and comprehensive vision of school mathematics offered by NCTM and on research about how students learn mathematics. (For information on the middle school curricula and links to Web sites dealing with elementary school and secondary school curricula, see the Show-Me Center Web site at showmecenter.missouri.edu.)

## EFFECT ON STATE STANDARDS AND ASSESSMENTS: THE EXAMPLE OF MISSOURI

The efforts of NCTM and NSF have significantly affected the composition of the states' curriculum frameworks and assessment programs. In Missouri, the Show-Me Standards and Curriculum Frameworks that emerged in 1996 were largely influenced by recommendations in the

NCTM *Standards.* To align state-sponsored testing programs with state and national standards, the Missouri Department of Elementary and Secondary Education replaced the Missouri Mastery Achievement Test with the Missouri Assessment Project (MAP) in 1997. The state's education leaders worked with committees of teachers, parents, and businesspeople to develop the new assessment in conjunction with CTB/McGraw-Hill, one of the nation's leading test publishers.

Designed to measure students' progress toward meeting new state standards, the MAP test differed from its predecessor in two important ways. First, while addressing such mainstay topics as number, algebra, and geometry, MAP also reflected the new emphasis on such mathematics content as probability, number sense, and discrete mathematics. Second, in addition to the familiar multiple-choice test format, MAP included short-answer, or constructed response, items and "performance events" items that required students to solve problems embedded in a real-world context, show their work, and explain their reasoning. See figure 17.2 for an example of eighth-grade test item.

> Each square inch of honeycomb contains approximately 25 cells. About how many cells would be found in a honeycomb that measures 8 inches by 12 inches? In the box below, provide the work that shows how you arrived at your answer

Fig. 17.2. An eighth-grade item from the 1998 MAP test

The new test departed from traditional assessment practices, and, consequently, many Missouri mathematics teachers and students struggled to adjust. Instead of valuing only the correctness of students' responses, MAP emphasized the problem-solving process. Suddenly Missouri students had to be able to read, interpret, and solve problems and communicate their problem-solving process to anonymous test graders. The changes in state testing programs added mathematical processes to the concepts that needed to be taught, thereby expanding the responsibilities of Missouri mathematics teachers.

For many Missouri teachers, the thought of making this transition was overwhelming. The state department of education addressed their concerns by implementing programs intended to ease the transition to the new state assessment. The NSF and mathematics teacher educators also furnished some unexpected help by offering Standards-based projects and professional development.

# STATE-LEVEL PROJECTS

AS STANDARDS-BASED mathematics curricula emerged from lengthy development and field testing, the NSF funded statewide projects to introduce these new curricula. In Missouri, the Middle Grades Mathematics Project was established for this purpose. This three-year professional development project introduced more than 100 middle school mathematics teachers and administrators to the four NSF curriculum projects that were completed or nearly complete at the time—Connected Mathematics, Mathematics in Context, Mathscape: Seeing and Thinking Mathematically, and Math Thematics. The participants not only studied the materials but also tried teaching some of the curricular units in their classrooms.

This experience enabled participating middle school teachers to adapt their teaching to such current state assessments as MAP. Working with these curricular materials, the teachers learned about integrated curricula and new mathematical content and became more aware of the kind of teaching required to promote problem solving, reasoning, and communication of mathematical ideas (Reys et al. 1998).

The Missouri Math Attack project also helped ease the transition from traditional to current assessment practices (Taylor, Campbell, and Long 2001). Initiated by the mathematics coordinator at the Missouri state department of education, the project was designed to "intensify the curriculum and activate instruction at the seventh-grade level." This project represented not only an effort to improve professional development but also an effort to implement the NCTM Standards through a focus on problem solving. Rather than have educators teach a traditional middle school mathematics curriculum—which was as much as 80 percent review (Flanders 1992)—Math Attack challenged teachers to teach only new topics. Topic reviews were integrated with the teaching of new material or embedded in the problems that students were solving. Integration allowed the teachers to focus on important mathematics concepts and help students make essential mathematical connections.

In their first few years of the Math Attack project, the teachers found themselves spending a significant amount of time rewriting the curriculum, designing lessons, and looking for activities. A tremendous payoff was realized, however, when their students were able to learn the "basics" in the process of solving interesting real-world problems. As a result of their experience with Math Attacks, the teachers came to believe that reiterating previously taught content and skills was not necessary. The results of basic-skills testing indicated that students from Math Attacks classes faired about the same as did students from traditional, review-laden classes. Moreover, the teachers in the Missouri Math Attack program reported that their students were not only learning mathematics concepts that were new and challenging to them but also becoming more competent problem solvers.

For teachers in the Missouri Math Attack project, the transition to the MAP test was made easier because they were already challenging their students to solve problems and reason mathematically. Therefore, they did not need to teach problem solving as yet another topic in an already crowded curriculum. As was the situation with the Middle Grades Mathematics Project classes, the students from Math Attacks classes were better prepared for the MAP test because they had been presented regular opportunities to communicate solution strategies in an integrated mathematics curriculum that was more focused than a traditional curriculum. As a result, the data showed that students of Math Attack teachers tended to outperform students from more traditional classes.

## MEETING STATE STANDARDS WITH INTEGRATED PROBLEM-SOLVING CURRICULA

HOW CAN middle school mathematics teachers comply with state standards and ensure optimal achievement for all students? One successful approach is to implement an integrated curriculum that focuses on problem solving. Instead of searching for a curriculum that specifically addresses each state standard as a major topic, teachers adopting Standards-based, integrated curricula often implicitly address the mathematical ideas and skills in their state frameworks. For example, teachers in the Missouri Math Attack project found that in presenting rich mathematical problems, they also addressed the basic skills that constituted a significant portion of the state standards. In essence, their students used basic skills in the process of solving larger problems. By solving nonroutine problems, the students found a reason to understand, recall, and apply basic skills. Similarly, teachers who have properly implemented the NSF-funded curricula described above have successfully met their state standards while covering *fewer* topics. Although some teachers perceive the need to furnish additional mathematics exercises, the spiraling nature of an integrated curriculum actually offers abundant opportunities for students to discuss, assess, and practice skills and procedures. This less-is-more approach is consistent with curriculum practices of such high-achieving countries as Japan, Korea, and Singapore. Moreover, the approach departs from typical U.S. eighth-grade mathematics classes, which, on average, cover nearly 50 percent more topics than do Japanese classes and which give only superficial treatment to topics (Peak 1996).

## USING AN NCTM STANDARDS–BASED APPROACH WITH A STATE STANDARD

The California state standard for sixth-grade geometry and measurement is as follows:

> **Grade 6—Geometry and Measurement**
>
> 1.3 Know and use the formulas for the volume of triangular prisms and cylinders (area of base × height); compare these formulas and explain the similarity between them and the formula for the volume of a rectangular solid. (California Department of Education 2001)

A cursory examination of this standard might lead teachers to develop a series of lessons designed to promote students' mastery of volume formulas. By using an integrated curriculum, however, this learning outcome can be covered several times and from multiple perspectives. Sixth-grade students in schools implementing the MathThematics curriculum (Billstein and Williamson 1999), for example, learn the formulas for volume of prisms and cylinders by exploring the Seven Wonders of the World. The volume of prisms emerges from the study of the Empire State Building, and the volume of cylinders is embedded in the study of ruins excavation at Mesa Verde, Colorado.

The connection between the formulas is made during the study of prisms and cylinders of the same height (Billstein and Williamson 1999, pp. 502–3). In this exploration, the students assemble two prisms and one cylinder, fill them with sand, and compute the volume of each one. The teacher then asks the students to describe common and distinguishing features of the three-dimensional objects and to describe a general method for computing volume. Finally, the students are asked, "How is knowing the formula for area of a circle important for finding the volume of a cylinder?" The *Math Thematics* teacher's guide (Billstein and Williamson 1999, p. 504) offers the following sample of a student's response:

> The bases of a cylinder are circles. You need to know the area of a base to find the volume of a cylinder, so you need to know how to find the area of a circle to find the volume of a cylinder.... The area of a circle is contained in the formula for the volume of a cylinder since $A = \pi r^2$ and $V = \pi r^2 h$ (or $V = B \cdot h$).

Such an investigation also helps the students understand that the formula for the volume of a cylinder is a natural extension of the formula for the volume of a prism. Such an approach not only enables students to develop general formulas and understand the relationships between them but also offers multiple opportunities for students to practice acquired problem-solving skills. Moreover, by incorporating the teaching of procedures into all lessons, the NSF *Standards*-based curricula ensure that mastery of isolated skills will not be compromised. Although formulas for prisms and cylinders are not a primary focus of any lesson in the subsequent two years of *Math Thematics*, their study is nevertheless embedded in other problems that are used to introduce new mathematics topics. Teachers using *Math Thematics*, as well as the teachers in the Math Attack program, have found success in an integrated, problem-solving curriculum that builds on the skills learned in previous years by introducing new and challenging mathematics instead of reviewing previously taught topics (Tetley 1998; Taylor, Campbell, and Long 2001).

The relationship between volume formulas is a prime example of the rich mathematical connections that can be made using an integrated problem-solving curriculum. Although the volume formulas are valuable components of the Seven Wonders lesson, educators should note that the unit does not list studying volumes as a primary objective. The mathematical relationship is embedded deeply within the curriculum and yet offers students a significant learning experience. Unfortunately, however, cursory examinations of the unit by textbook adoption committees, and indeed many others, could lead to the erroneous conclusion that the unit does not adequately meet state curriculum standards.

## CONCLUSION

IN THE United States, most of the nearly 16,000 school districts design their own curricula, usually within the broad guidelines of their state's curriculum framework or standards (Peak 1996). Each district or state adopts a mathematics textbook that covers its framework. Typically, many educators perceive the textbook selection process as one that begins by examining a set of books. The process, however, actually originates with determining the set of textbooks to be considered. A pivotal decision, therefore, is to determine which characteristics the school district values most. Then the school district can include in the review process the textbooks that embody those values. School districts that adopt an integrated curriculum focused on problem solving give their teachers curricular materials that are consistent with NCTM's vision of school mathematics. Such curricula genuinely engage students in learning sound and significant mathematics. Because traditional middle school mathematics curricula dedicate 80 percent of their coverage to review of topics, they do not afford such opportunities for substantial learning.

Through the previous three *Standards* documents and the recently released *Principles and Standards for School Mathematics* (NCTM 2000), NCTM offers classroom teachers, school districts, state departments of education, and textbook publishers continued guidance toward its vision of a coherent, focused mathematics curriculum. The challenge for school districts and teachers, then, is to use that guidance carefully as they examine their mathematics curricula options. To fairly and accurately examine integrated, problem-solving curricula, teachers must review the materials, try out entire units in their classrooms, and report to one another. Such collaborative curriculum investigation not only can lead to selecting a curriculum that will meet state standards but also can contribute to the professional development of the teachers (Reys et al. 1997).

## REFERENCES

Billstein, Rick, and Jim Williamson. *Middle Grades MathThematics*, Book 1. Evanston, Ill.: McDougal Littell, 1999.

California Department Education. *Mathematics Content Standards for California Public Schools: Kindergarten through Grade Twelve.* 2001. www.cde.ca.gov/board/pdf/math.pdf. (4 June 2002)

Education Development Center. *MathScape: Seeing and Thinking Mathematically.* Columbus, Ohio: Glencoe/McGraw-Hill, 1998.

Flanders, James. "Textbooks, Teachers, and the SIMS Test." *Journal for Research in Mathematics Education* 25 (May 1994): 260–78.

Kilpatrick, Jeremy. "A History of Research in Mathematics Education." In *Handbook of Research on Mathematics Teaching and Learning*, edited by Douglas A. Grouws, pp. 3–38. Reston Va.: National Council of Teachers of Mathematics, 1992.

Lappan, Glenda. "Revitalizing and Refocusing Our Efforts." *Journal for Research in Mathematics Education* 30 (November 1999): 568–78.

National Council of Teachers of Mathematics (NCTM). *Curriculum and Evaluation Standards for School Mathematics.* Reston, Va.: NCTM, 1989.

———. *Professional Standards for Teaching Mathematics.* Reston, Va.: NCTM, 1991.

———. *Assessment Standards for School Mathematics.* Reston, Va.: NCTM, 1995.

———. *Principles and Standards for School Mathematics.* Reston, Va.: NCTM, 2000.

Peak, Lois. Pursuing Excellence: *A Study of U.S. Eighth-Grade Mathematics and Science Teaching, Learning, Curriculum, and Achievement in International Context.* Washington, D.C.: National Center for Educational Statistics, 1996.

Reys, Barbara, Robert Reys, David Barnes, John Beem, and Ira Papick. "Collaborative Curriculum Investigation as a Vehicle for Teacher Enhancement and Mathematics Curriculum Reform." *School Science and Mathematics* 97 (May 1997): 253–59.

Reys, Robert, Barbara Reys, David Barnes, John Beem, and Ira Papick. "What Is Standing in the Way of Middle School Mathematics Curriculum Reform?" *Middle School Journal* 30 (December 1998): 42–48.

Silver, Edward. *Improving Mathematics in Middle School: Lessons from TIMSS and Related Research.* Washington, D.C.: U.S. Department of Education, 1998.

Taylor, P. Mark, Larry Campbell, and Vena Long. "Math Attack: Balancing the Equation for Middle School Mathematics." *Middle School Journal* 32 (January 2001): 36–43.

Tetley, Linda. "Implementing Change: Rewards and Challenges." *Mathematics Teaching in the Middle School* 4 (November 1998): 160–67.

18

# A Teacher's Reflection on Using an Integrated Approach to Teach Mathematics

Lecia Bentley with Anita Bowman

UNTIL I started writing this chapter, I never thought about labeling my sixth-grade mathematics instruction as "integrated mathematics." I know that I teach mathematics differently than I did when I began teaching sixteen years ago, and I am aware that my current mathematics textbook, *Middle Grades Math Thematics* (Billstein and Williamson 1999), is different from books that I have used in the past. Recently, however, as I reflected about my practice and read about integrated mathematics, I began to realize that I have been using an integrated approach for years.

## How My Mathematics Instruction Reflects an Integrated Approach

INSTEAD OF building instruction around units (e.g., a textbook chapter on addition and subtraction of fractions), my instruction is organized in modules. My textbook features eight modules of five or six sections each. In the course of each module, students actively participate in a variety of activities that help them connect important mathematical ideas. In one module, for example, the six sections focus on developing students' mathematical reasoning skills and conceptual understanding in several mathematical areas—probability, divisibility, decimal and fraction multiplication, multiples and mixed numbers, and equations and graphs. Hence, the mathematics content is integrated within the modules.

Mathematics content is also integrated within the individual sections of the modules. For example, in one section, as students play games involving theoretical and experimental probability, they are connecting numeration concepts related to fractions and decimals and practicing converting fractions to decimals and decimals to fractions. They are learning some benchmark conversions (e.g., 3/4 = 0.75), and they are learning to decide when using a calculator is more expedient than paper and pencil for making conversions.

The integrated nature of mathematics also comes through in written assessments. For example, figure 18.1 shows a writing task that I assigned and the responses of one of my students. I think such writing exercises help students connect mathematical procedures and concepts.

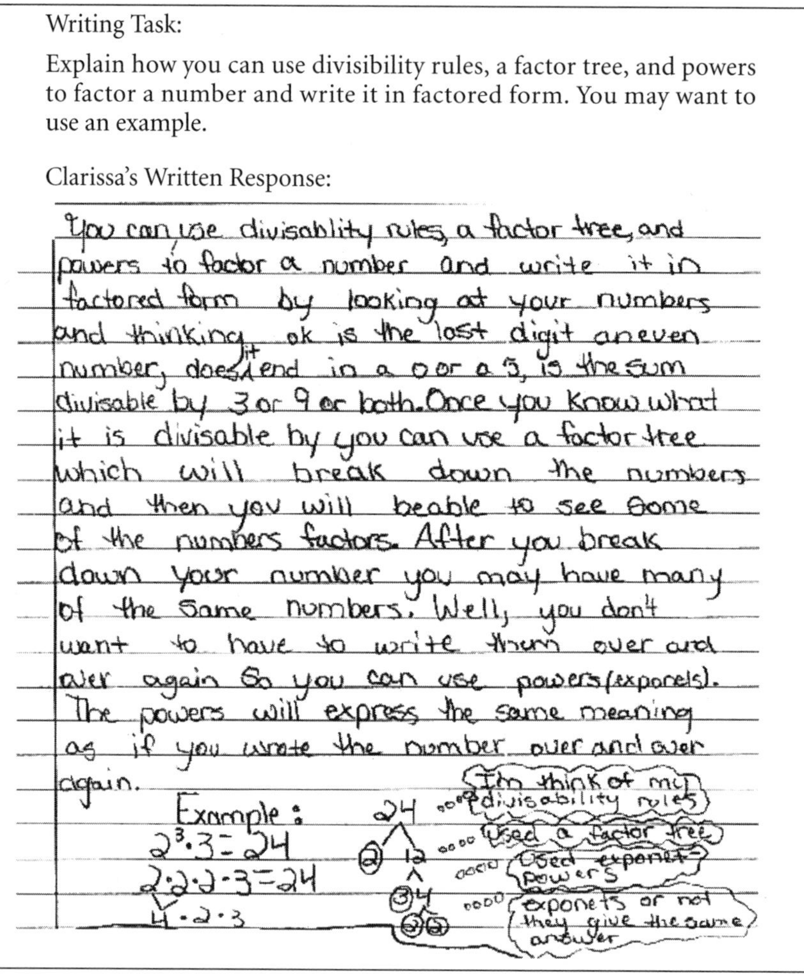
Writing Task:

Explain how you can use divisibility rules, a factor tree, and powers to factor a number and write it in factored form. You may want to use an example.

Clarissa's Written Response:

You can use divisablity rules, a factor tree, and powers to factor a number and write it in factored form by looking at your numbers and thinking ok is the lost digit an even number, does it end in a 0 or a 5, is the sum divisable by 3 or 9 or both. Once you know what it is divisable by you can use a factor tree which will break down the numbers and then you will beable to see some of the numbers factors. After you break down your number you may have many of the same numbers. Well, you don't want to have to write them over and over again so you can use powers (exponets). The powers will express the same meaning as if you wrote the number over and over again.

Example:

$2^3 \cdot 3 = 24$

$2 \cdot 2 \cdot 2 \cdot 3 = 24$

$4 \cdot 2 \cdot 3$



Fig. 18.1. An example of a writing task and a student's response

Good mathematical understanding results from developing big mathematical ideas by connecting many smaller ideas, procedures, and memorized facts. I try to give students opportunities to develop these larger structures through reflective writing and classroom discussions. However, for powerful mathematical ideas to develop, writing about and discussing mathematical ideas need to take place within mathematically rich, integrated activities.

Integrating mathematics requires more than just integrating content within mathematics. Because I teach both science and mathematics to the same groups of students, I look for opportunities to integrate the concepts from these subjects. Sometimes I use such resources as AIMS (Activities Integrating Mathematics and Science) activities (see www.aimsedu.org) to integrate mathematics and science topics.

Even when I am teaching only mathematics, I think of integrating in several ways. Within my classroom, I incorporate the process skills of problem solving, mathematical reasoning, and communication in almost every lesson. Furthermore, such processes as the representations of mathematical ideas serve as important vehicles for integration. For example, when my students are learning about powers of numbers, I make sure that they see powers in different representations (e.g., as folded paper, as symbols in expanded form, as symbols in exponential form, and as factor trees) and that they understand how the various representations relate to the same main concept.

## Development of My Mathematics Instruction

I entered the teaching profession in January 1985 after completing my intermediate education degree and obtaining state teaching certification in general education (grades 4–6) and mathematics (grades 6–9). In college, I decided that I wanted to teach middle school mathematics because of my own experiences as a middle school student. I had attended kindergarten through grade 8 in a school that did not offer first-year algebra to eighth-grade students. For me, mathematics in these years consisted of computation; students repeatedly performed operations on whole numbers, decimals, and fractions. Then in high school, when I faced algebra 1—my first mathematics course that did not focus on endless computations—I struggled. I continued to struggle through college mathematics courses, and I was still trying to learn mathematics content during my mathematics methods course.

For my first two years of teaching, I taught seventh- and eighth-grade students. I based my instruction on a traditional textbook organized around units. At that point, I considered the mathematics instruction that I delivered better than that which I had experienced as a student, because at least my students were studying geometry, statistics, and probability as well as prealgebra concepts.

From August 1988 through June 1990, I taught eighth-grade mathematics and one section of seventh-grade prealgebra at the Saudi Arabian International School in Riyadh. The next year, I taught sixth-grade mathematics until I left the country in February 1991 during the Gulf War. While teaching in Saudi Arabia, my principal gave me "free rein" regarding how I taught the required mathematics curriculum. The mathematics department chair was a very creative person who added many arty touches to her classroom and used multiple resources in her teaching. Even though she taught geometry, she used activities from algebra. We often used activities from the *Arithmetic Teacher;* the number of mathematical concepts that surfaced as our students completed those activities sometimes surprised us. Our superintendent traveled to the United States about the time that NCTM published *Curriculum and Evaluation Standards for School Mathematics* (1989), and he brought us copies of the document. We regularly attended the Near East and Southeast Asian Mathematics Conferences where the focus was on Standards-related mathematics instruction. I guess we were, in a sense, a mathematics department "ahead of our time."

When I returned to the States, I quickly found myself teaching middle school mathematics again. One of my first assignments was teaching *Transition Mathematics* (University of Chicago School Mathematics Project 1992, 1998)"—essentially a prealgebra course with geometry and algebra integrated throughout. I taught from *Transition Mathematics* with my higher-level students and taught regular mathematics to two other groups of students. The textbook for the regular mathematics curriculum was quite traditional, and I never used it. Sometimes I could use an activity from the *Transition Mathematics* book with my regular students. Other times I would use activities from other instructional resources. Finding activities took a lot of time, but I felt comfortable with the approach. I knew the state-defined curriculum for my grade level and therefore the mathematics that my students needed to learn. I kept folders for each state objective and filed related learning activities as I came across them.

In 1993, I was asked and agreed to pilot-test the integrated mathematics materials developed by the National Science Foundation (NSF)–funded middle school curriculum project known as the Project STEM (Six through Eight Mathematics) project. The project interested me because all the activities were organized and ready to be used, reducing my need to constantly search for instructional activities. Furthermore, I had spent the previous two years in an NSF-funded Calculator Institute designed to help middle school teachers offer better mathematics instruction through the use of handheld calculators. The calculator project emphasized increased knowledge of mathematics content and understanding of reform mathematics. When I went to the two-week summer training for the STEM pilot in summer 1993, I realized that I had a better understanding of the mathematics content and a better sense of *Standards*-based mathematics than most of the other trainees. Most of the teachers had three main concerns about teaching this integrated coursework: (1) they were unsure how to manage integrated instruction in the classroom, (2) they wondered where they would get the materials (i.e., manipulatives) called for in the activities, and (3) they worried that they would not be able to answer students' questions because the teachers were not sure that they themselves understood the mathematics content well enough. Because I had already been teaching with an integrated approach, I felt totally comfortable with pilot-testing the STEM materials.

After the 1993–94 pilot year ended, I requested permission to continue teaching with the materials, and my school purchased the materials in draft form. For the next few years, I continued to use some of the STEM activities while still pulling activities from other sources. I did not use the traditional textbooks furnished by the school district; those textbooks did not support good NCTM Standards–based mathematical instruction. Beginning in August 1999, my district adopted the STEM materials for grades 6–8, now known as *Math Thematics* (Billstein and Williamson 1999), and I have been using these materials for the past few years.

## Challenges and Issues Involved in Integrated Mathematics Approaches

Integrated mathematics instructional materials and teaching approaches are very different from traditional materials and approaches. To move my mathematics instruction from a traditional approach to an integrated one, I had to—

1. change my perceptions of my role and my students' role in the classroom;

2. pay special attention to my management of materials and time as well as to the students' behavior;
3. work within my school environment by ensuring that what I did in my classroom was supported, or at least tolerated, by my principal and the other teachers;
4. communicate effectively with parents about what an integrated approach to mathematics would mean for their children's learning;
5. ensure that I addressed all objectives in the state-defined mathematics curriculum for my grade level; and
6. attend to achievement, assessment, and testing issues.

## CLASSROOM ROLES

IN AN integrated mathematics classroom, I serve mainly as a facilitator. I begin the lesson, and then I keep the students moving—not by showing them how to do the task but by asking questions to help them focus on its important aspects. Sometimes the students want me to solve the problem, and I even catch myself starting to do so, but then I stop and think, "It won't help them learn mathematics if I do the task for them." At that moment, I give the task back to the students, but perhaps with some helping hints.

The level of students' participation requires a close watch. Some students are reluctant to participate. For example, half the school year passed before one of my shier students would voluntarily contribute to our class discussion. An environment based on integrated materials and activities can allow some students to dominate the class; as a teacher, I must keep working to encourage the quieter and more hesitant children to participate.

In a Standards-based integrated mathematics classroom, I know that I need to understand the mathematics content well so that I can assess my students' understanding. I check their understanding through their written work and oral explanations and their contributions to classroom discussions. Using integrated mathematics materials with my students has helped me increase my own understanding of mathematics content. I have also learned more about how students come to understand certain mathematical concepts and how to help students overcome common mathematical misconceptions. Published research on how students internalize mathematical ideas has been very helpful to me.

The teacher's role in an integrated classroom requires careful listening and questioning. I must listen closely to comprehend what my students are saying. I am constantly asking the students "How did you get that answer? What are you thinking? Explain. Tell me more."

At the same time, the students must take bigger roles in communicating with one another. I teach my students how to communicate mathematically. I emphasize mathematics vocabulary. At first, I let the students explain the mathematics to me in their own words, but then I push them to use proper vocabulary. For example, a student might say, "I multiplied by the second number," and I will ask, "What do we call that number?" and will prompt the student for the correct vocabulary (*factor*). For each module, I post mathematics vocabulary words around the classroom. As the students discuss mathematics, I ask them to look around for a more precise word. Many times, I will not continue until the students rephrase their explanations in mathematical terms. I consider their initial explanations to be important "first drafts," but learning also requires that they edit their explanations as they reflect on their thinking; otherwise, the students do not learn to talk about mathematics.

My students have adjusted well to the way I teach. They like being actively involved in explorations, working in pairs or groups, and having opportunities to talk about their mathematical ideas. At the end of the school year, I asked my students to write letters to the fifth-grade students who would be in my class the next year. Here are some samples of what they wrote:

- "I liked working in groups or pairs, I liked playing games to learn the math, and I liked learning and seeing the vocabulary."
- "The mathematics book is usually telling stories, legends, and sometimes even jokes. Usually the stories and legends and jokes give me an idea of what we are about to learn or get into."
- "Mathematics, in my opinion, is the most interesting class you will ever take.... I think it makes you feel good when your parents are doing paperwork, bills, or taxes, and you know what all the numbers mean and you can help them out. I have liked math the best this year, even though I did not get all A's in math; but it was a challenge for my mind learning new things."
- "When I was in the fifth grade, I hated math because I thought it was hard but, now, I like math and I think [that the new sixth-graders] will love to do this. I want to learn more about math."

- "I loved sixth-grade math because of the many different things we did. We did many things I had never done in math before. We also did some things that I had done, but we looked at those things in different ways. It helped me feel so much more confident when it came to end-of-grade testing."
- "In sixth grade, math is easy but also hard sometimes. In sixth grade, math is sometimes very annoying. You have to show how you got certain answers a lot. Teachers might even give you cubes or a game to find things out. You may have to keep up with a notebook but you get to do graphs that end up looking like funny pictures and play games with dice and blocks and that is why math is fun. You might also have to keep up with a portfolio with your work."

I find it encouraging that my students talk about liking mathematics even while they agree that mathematics "is sometimes hard." By the time we come to the end of a school year, I see dramatic improvements in the children's understanding of mathematics; ability to apply mathematics in real-world, problem-solving activities; confidence in their ability to do mathematics; and enthusiasm for, and attitude toward, mathematics.

One serious problem that I encounter with my teaching approach involves absences. When students miss class, I cannot just send the work home for them to make up. When they are not part of the classroom activity and associated discussion, they are not learning the mathematics. An isolated absence is not a problem, but in the past, I have had students whose multiple absences have affected their progress.

## Management of Materials, Time, and Students' Behavior

FOR THE STEM pilot and the next couple of years, I worked with inner-city children. I had to prepare all materials ahead of time so that I could get the students working on task immediately. Before the lesson, I would place each group's materials into separate bags. Later, I purchased clear plastic boxes for storing items by type and baskets for holding materials for the day's lesson. Then I would assemble the baskets for a particular activity and place the materials back into the boxes at the end of class. This system has continued to work well for me. Because behavior can be a problem, I attend to the ways I group students. But I have found that if I have planned a lesson that is meaningful, engaging, and fast-paced, the students' behavior is usually good.

Because students are actively involved in mathematics activities during class, timing and planning require special attention. Meeting learning objectives sometimes does not fit nicely into a forty-five-minute or one-hour class. For example, I ran out of time in a recent class because I let my students play a game longer than I had initially planned; they were engaged in important ideas that they needed to learn, so I chose to let them continue with the activity. Long-term planning is also essential. I have to plan carefully for us to cover all eight modules by year's end. To complete our curriculum, I push daily for my students to work efficiently. If they are not working diligently, I walk around the room making notes. My close observation sometimes intimidates them, but it also makes them focus. I also reward the group that works best or the group that has the best discussion. I have found that with a new class, I reward them a couple of times, and then I do not have to do it again because they understand what I expect of them. The students also quickly learn to ask one another questions similar to the ones that I have asked them.

I want students to write about mathematics but cannot spend a lot of time letting them write by themselves. Sometimes I limit the writing and hold discussions. Students often learn to communicate their understanding just as well in discussion as they do in writing. I make frequent transitions between individual work, group work, and class discussions. Those transitions are not easy. Some days, I get very frustrated and think, "Today, the students are going to do seatwork."

## THE SCHOOL ENVIRONMENT

MY SCHOOL principals have been generally supportive. When I pilot-tested the STEM materials, my principal gave me permission to use the materials as long as I covered the state objectives. When he visited my classroom, he was impressed with my transitions between instructional methods, which kept the students moving and linked concepts within lessons. He sent other teachers to observe my class. I think that my principal would question my teaching approach only if my students' end-of-grades test scores fell.

In the past, some of my colleagues did not want to use Standards-based integrated mathematics materials, but they were not concerned that I used them. Before I piloted the STEM materials, I was concerned about working in isolation in a style that was very different from that of the other sixth-grade mathematics teachers—even though I knew that my curriculum met state objectives and that my students were learning mathematics. NCTM *Curriculum and Evaluation Standards for School*

*Mathematics* (NCTM 1989) and its accompanying grade-level Addenda series, as well as *Mathematics Teaching in the Middle School* articles, reafirmed for me that my teaching approach was sound. However, having the company of other teachers would have been of help to me.

With the district's recent adoption of the *Math Thematics* materials, I suddenly found myself surrounded by five colleagues also using Standards-based integrated-mathematics materials. One is a veteran teacher who began using integrated materials in 1994 but has not been trained on the materials or in using manipulatives to teach mathematics. She tends to mix integrated and traditional teaching. For example, if she is teaching an integrated activity and realizes that some of the students are confused, she will change to a more traditional teaching mode. I try to support all my colleagues as they implement the materials. We use grade-level planning time to collaborate. I like to stay a little bit ahead of them in the textbook so that I can share with them some experiences to watch for as their students study the modules. For example, when my colleagues prepare to teach the second half of a module, we talk about what the students might do, what content should be emphasized, how the activities fit into the state curriculum, and what activities the teachers might omit if time is limited; frequently we meet again to address these same issues when their classes have worked about halfway through the module. Of course, the teachers worry about their students' end-of-grade test scores. To reassure them that their students will do fine, I point out the objectives that the teachers have covered in the various modules. I also encourage them to keep moving because teachers new to integrated curricula tend to get bogged down if students have not yet mastered the mathematics content.

## PARENTS

A FEW parents have had trouble adapting to the integrated approach that I use, especially when I was pilot-testing the STEM draft and did not have a textbook to send home. In that situation, those parents generally had difficulty identifying the mathematics in their children's work. Now that my school has the published book in which the mathematics content is highlighted in each section, such parental confusion is less of an issue. When the parents see the mathematics, they also see how substantial it is. A parent once told me, "This math moves too fast; *I* can't keep up." The parents who voice concerns are generally those who will sit down with their children and talk about the homework. They sometimes tell me that they think the lessons are tough, but they do not see that difficulty as a problem, because they want instruction to be rigorous and they want to see learning take place.

## The Curriculum

I know my state curriculum and can identify the curriculum content in the activities that my students do. I do not actually keep organized records; rather, I mentally monitor what content is being taught. I can keep mental tabs because I have used these materials for several years. A checklist would probably be helpful to someone who has just started teaching integrated mathematics.

I also find that substantial long-range planning is needed to ensure that integrated mathematics materials meet all the objectives within state-defined curriculum.Teachers must also know their state's objectives and ensure that their curricula meet those objectives. When my district adopted integrated mathematics materials, I worked on the district's nine-week pacing guides. As we researched the guides, we found one objective from the state curriculum that was not addressed in the textbook; therefore, we have supplemented the textbook with an activity designed to address that objective.

## Achievement, Assessment, and Testing

During my first two years of teaching, I used a traditional textbook and a traditional approach. I taught for mastery. I had a checklist for each objective or skill in the state-defined curriculum, and I would mark off each one as the students showed that they had mastered it. Then, at the end of the year, I discovered that my students could no longer apply the skills that I had checked off.

I do not focus on mastery learning now. Because my instruction is based on integrated units, my students revisit important mathematical ideas several times and in several ways throughout the year. As a learner myself, I find that such an approach makes sense. I know that I do not always fully understand a concept the first time around, but if I see it again and think about it, I can grasp it. Similarly, if I deem a concept not important or related to me, I often do not make sense of it. I think that those same processes happen with my students. For example, the first time the students encounter addition of fractions, they might not learn as much as I had hoped, but later, when they want to play a game that requires them to add fractions, they might engage in learning more about how to add fractions because it is then useful to them. For the first couple of years that I used an integrated-mathematics approach, I would become frustrated around the beginning of December because my students did not appear to be learning mathematics. I knew that I had been "planting seeds," but I saw no crop to show for my efforts. Then, by about February or March, I would suddenly see how much that they had learned. I did not believe it the first time that I saw their knowledge bloom. The Standards-based, integrated approach truly improves students' achievement, but the approach has caused me to rethink my timetable for students' development of mathematical understanding.

Assessment in my classroom is generally informal. I try to determine what mathematics my students are learning. I might ask them to write or say more about a specific problem, or I might use a problem to spot-check for understanding. If I collect homework, I sometimes just focus on the explanation that the students give for the solution of a single problem. Sometimes I give a little quiz, but I do not give big tests. I consider many factors in determining grades (e.g., the extent to which the students are participating in tasks and discussions, their demonstrated insights about mathematics, and so forth). Most of my assessment activity helps me decide what direction my instruction should take. When I know what my students do and do not understand, I can decide to do more work with the concept or move on if I know that they will have another opportunity to develop the concept more fully later. Sometimes when I introduce a task, I assess students' knowledge of the underlying content, and sometimes I have to prompt them to think about what they already know. For example, I might ask the students to talk about how to change fractions to decimals before they begin playing a probability game that requires that skill.

My state requires a high-stakes test, the results of which have ramifications for the students, teachers, principals, and district-level administrators. Testing does not change my perception of good mathematics teaching, but I certainly am concerned about the end-of-grade testing results, even though my students have always done well in the past. On any given day, the students may come in and perform miserably on a test. I do not have much control over such circumstances. I might believe that I have prepared my students well, but their test scores are not necessarily going to be high as a result. However, I know that the methods I use and the content I teach are aligned with the state-defined curriculum, which, in turn, is aligned with the NCTM *Standards*, including *Principles and Standards for School Mathematics* (NCTM 2000). I also know that my state's end-of-grade tests are aligned with the state curriculum.

Teachers in my state usually spend the six weeks leading up to the test reviewing and preparing students. Until last year, I, too, would stop teaching textbook materials about two weeks before the test and use state review materials until the test. Last year, my school had a special guided-studies class that we used to prepare students for testing, so I did not have to interrupt the students' usual mathematics instruction.

## ADVANTAGES OF AN INTEGRATED APPROACH

MIDDLE SCHOOL students bore easily. Even as a teacher, if I taught only decimals day-in and day-out for six weeks, I would be bored, too. I think Standards-based integrated mathematics works because it keeps students' attention better than a traditional approach does.

Middle school students need opportunities for socialization. Within the group work that is an important part of the integrated approach, the students learn with concrete materials, and they talk about what they are doing. Not all the students' discussions are about important mathematics; sometimes the discussions revolve around learning to share, listen, and respect one another.

Students at this level also need to be challenged. With an integrated-mathematics approach, I challenge students with significant problems that require that they learn and apply a variety of mathematical concepts. Within their groups, they have opportunities to explain their thinking about the problem and to learn how other students are thinking. Often an idea expressed by one student challenges other students to rethink their ideas. My students know that whenever they write down an answer, they need to be prepared to explain their answer to me.

Using good integrated-mathematics materials does not mean covering a concept once and leaving it behind. Instead, it means that teachers and students come back to the same mathematical concepts and skills several times during the year but in different contexts and in different ways. Because of this approach, the students can connect important mathematical concepts and develop "big ideas." Sometimes I will hear a student say something like, "This is like the handshake problems that we did a couple of months ago." When a student can connect the handshake problem with the problem of scheduling a single-elimination soccer tournament and then with a consideration of a geometric pattern of triangular numbers, then I know that the student is developing important mathematical ideas.

## RECOMMENDATIONS TO TEACHERS NEW TO INTEGRATION

FOR TEACHERS trying to switch to a Standards-based integrated mathematics approach, I recommend that they just jump in and plow through. Initially, it is easy to get discouraged and regress to the traditional approach. For example, last year—the first year all middle school teachers in my district were required to use integrated-mathematics materials—several teachers gave up part way through and began teaching from the old textbooks. However, the teachers who did not give up are glad. This year, they see more of the possibilities for students' learning and see students enjoying mathematics. They are also seeing higher test scores.

Those who are new to teaching integrated curricula should try to collaborate with another teacher. Even if the other teacher is also new to the approach, teaching partners can help each other simply by sharing what is happening in their mathematics classrooms. Talking with other teachers who have been teaching integrated curricula for a while is especially helpful; often, their perspectives can make new teachers' transitions easier.

The NCTM *Standards* materials help tremendously. If integrated mathematics is considered in the context of the *Standards* documents, the approach clearly seems well grounded in the movement toward higher-quality teaching and learning. I suggest that teachers new to the integrated mathematics approach might use *Standards* materials to help them gain administrative and parental support for their use of the materials. Also, the *Standards* materials might be used as a teaching resource because they contain many activities that could be used as supplements to a particular set of integrated mathematics materials.

Lastly, I encourage those new to teaching with integrated mathematics materials not to be afraid to thoughtfully experiment with their mathematics instruction. I think they will be pleased with the results—I know I have been. My students' end-of-grade test scores are significantly higher than their scores for the previous year. More important to me, I find that my students can really see the different mathematics within problem-solving tasks. They are able to pull problems apart and analyze the pieces. When they begin a problem-solving task, they are confident that they can get to one or more reasonable solutions because they have developed excellent reasoning skills. They are enjoying mathematics, and because they enjoy it, they are more consistently, and more deeply, engaging in mathematical thinking—and this increased engagement is leading to substantial increases in learning.

## REFERENCES

Billstein, Rick, and Jim Williamson. *Middle Grades Math Thematics*, Book 1. Evanston, Ill.: McDougal Littell, 1999.

National Council of Teachers of Mathematics (NCTM). *Curriculum and Evaluation Standards for School Mathematics.* Reston, Va.: NCTM, 1989.

———. *Principles and Standards for School Mathematics.* Reston, Va.: NCTM, 2000.

University of Chicago School Mathematics Project (USCMP). *Transition Mathematics.* Glenview, Ill.: Sott Foresman Addison Wesley, 1992, 1998.

# 19

# An Example of Districtwide Adoption and Implementation of Integrated Mathematics Curricula

Everly Broadway with Anita Bowman

FOUR YEARS ago, our school district began implementing integrated mathematics curriculum materials for kindergarten through twelfth grade. We chose to phase in the curriculum for K–8 over a three-year period. At the high school level, we chose to phase in the integrated curriculum alongside the traditional sequence of first-year algebra, geometry, and second-year algebra. In this chapter, I briefly describe our school district, Durham Public Schools in Durham, North Carolina, and our NSF-funded Local Systemic Change project. I then discuss how the district prepared for the project. I subsequently highlight the challenges and opportunities that we have faced and how we have dealt with them. I also describe ongoing advocacy work in support of the project.

## THE SETTING FOR CHANGE

In 1999, with major funding from the National Science Foundation (NSF) under its Local Systemic Change through Teacher Enhancement in Mathematics (LSC) initiative, our school district embarked on a journey to dramatically change the teaching of mathematics districtwide. The four-year, NSF-funded (NSF#9819542) project titled Realizing Achievement in Mathematics Performance is locally referred to as Project RAMP. While continuing as the mathematics coordinator for our district, I have also been serving as one of the principal investigators for the project.

The curricular basis for Project RAMP is materials produced by three of the thirteen integrated mathematics projects funded by NSF in the early *1990s—Investigations in Number, Data, and Space* for kindergarten through grade 5 (TERC 1997), *Mathscape: Seeing and Thinking Mathematically* for grades 6–8 (Education Development Center 1998), and *Contemporary Mathematics in Context* for grades 9–12 (Core-Plus Mathematics Project 1998). (See chapter 4 in this book for brief descriptions of these and the other ten NSF-funded curriculum projects.) In general, the instructional materials produced by each of the thirteen curriculum projects were designed around the framework established by *Curriculum and Evaluation Standards for School Mathematics* (NCTM 1989); thus, each of the curriculum projects paid special attention to the processes of problem solving, communication, reasoning, and mathematical connections. The mathematics section of our statewide mathematics curriculum is also based on the 1989 *Standards* and the more recent *Principles and Standards for School Mathematics* (NCTM 2000). Because the curriculum projects and the state mathematics curriculum share a common base, the instructional materials that we have selected align well with our state-defined curriculum.

## DESCRIPTION OF THE SCHOOL DISTRICT

DURHAM PUBLIC Schools, which serves children in the entire county of Durham, North Carolina, was formed in 1992 from the merging of two systems, one primarily urban and the other primarily suburban and rural. Currently, our district is the sixth largest public school district in the state, with approximately 30,546 students in prekindergarten through twelfth grade. Our student ethnic distribution is 56.1 percent African American, 30.7 percent white, 7.8 percent Hispanic, 2.4 percent Asian, 2.7 percent multiracial, and 0.3 percent American Indian. Approximately 39.4 percent of our students qualify for free or reduced price lunch, although the rate ranges from 11.3 percent at one elementary school to 96.1 percent at another elementary school. In the 2000–2001 school year, 513 students dropped out of Durham Public Schools, compared with 731 students the previous year. Although this trend is good, it is still true that students who drop out are disproportionately African American, constituting 70.6 percent of the total dropouts.

In 1995, a statewide school accountability program was initiated. This program, titled the ABCs of Public Education, established systems of rewards and consequences for students, teachers, principals, and central services administrators. In this program, the results of high-stakes achievement tests are used to assess schools on how well they are meeting their goals. These high-stakes tests are referred to as end-of-grade (EOG) and end-of-course (EOC) tests (e.g., the third-grade mathematics EOG test or the algebra 1 EOC test). These tests are used to measure growth. Schools are labeled according to their results (e.g., exemplary growth/gain, expected growth/gain, and low performing). Table 19.1 shows the district schools' distribution based on the state's *ABCs of Public Education Report 2001* (North Carolina Department of Public Instruction 2001). Durham Public Schools has five elementary schools, one middle school, and one high school labeled as exemplary growth/gain. One high school is labeled as low performing. Fortunately, the ABC mathematics testing program and state mathematics curriculum are well aligned.

TABLE 19.1. ABC Status by School Type

| | School Type | | | |
|---|---|---|---|---|
| ABC Status | Elementary | Middle | High | Special |
| Exm[1] | 5 | 1 | 1 | 0 |
| Exp[2] | 15 | 3 | 3 | 3[5] |
| NR[3] | 4 | 5 | 0 | 0 |
| LP[4] | 0 | 0 | 1 | 0 |

[1]Exm—Exemplary Growth/Gain

[2] Exp—Expected Growth/Gain

[3] NR—No Recognition

[4] LP—Low Performing

[5] 1 Special 6–12 Magnet School; 2 Special Alternative 6–12 schools

Source: *2000–2001 ABCs Results*

Within our community, several highly vocal advocates speak up for special groups of students (e.g., underachieving students, gifted students, and students in particular neighborhoods). Our district is trying to help these advocates and the community as a whole begin thinking about focusing on the high-quality work and continuous progress that will create a climate in which all students can succeed. The district's staff and the community members agree that the disparity in mean scores when achievement data are disaggregated by ethnicity and special classifications is a great concern.

At the end of the first year of Project RAMP, on the1999–2000 fifth-grade EOG mathematics test, 69.3 percent of African American students and 94.9 percent of white students scored at or above grade level, and on the 1999–2000 algebra 1 EOC test, 37.7 percent of African American students and 75.5 percent of white students scored at or above grade level. At the end of the third year of project RAMP, on the 2001–2002 fifth-grade EOG mathematics test, 83.4 percent of African American students and 96.7 percent of white students scored at or above grade level, and on the 2001–2002 algebra 1 EOC test, 63.5 percent of African American students and 85.6 percent of white students scored at or above grade level. All stakeholders are willing to work to help close these achievement gaps, and teaching mathematics using integrated mathematics materials is one of the district's strategies for closing the racial achievement gap in mathematics.

## DESCRIPTION OF THE PROJECT RAMP PROGRAM

THE FOLLOWING three goals guide the NSF funding of Project RAMP:

1. To provide comprehensive staff development for all mathematics teachers in kindergarten through twelfth grade as they implement an instructional program and materials consistent with the NCTM Standards.

2. To create an infrastructure that supports ongoing, collegial professional development in mathematics education.

3. To build administrative and community support for continuous improvement in mathematics education.

Because professional development lays the essential foundations for meeting these goals, Project RAMP supports all 980 teachers of mathematics in kindergarten through twelfth grade by offering more than 100 hours of ongoing professional development in mathematics pedagogy and content over a four-year period. Professional development providers include Project RAMP staff, publishers' consultants, school-based teacher leaders, and university-level mathematics education and mathematics faculty. The formats for teacher professional development include districtwide grade-level and course-level workshops, monthly school-based study groups, and classroom-level coaching. All professional development sessions are designed to help teachers increase their understanding of curriculum organization, mathematics content, assessment, and planning. Because most of the district's teachers originally learned mathematics in traditional settings, a major objective of professional development in Project RAMP is to help teachers understand important connections among various strands of mathematics and to help teachers learn to attend to, and build on, the student-realized connections that arise in their classrooms during instruction. Project RAMP further supports the high-quality teaching of Standards-based mathematics curricula by conducting workshops for principals, district administrators, parents, and the community at large.

Our district's vision for mathematics instruction in kindergarten through twelfth grade is that all teachers offer high-quality mathematics instruction. We will know that this vision has been realized when all students are successfully learning challenging mathematics content in inviting and exciting learning environments with appropriate materials. Furthermore, all administrators at both the district and school level will be supporting the teaching and learning of mathematics with effective leadership, teacher recruitment and evaluation, necessary policies, and sufficient budgetary resources. Most important, the organizational culture will be supporting continuous improvement for teachers, students, and administrators and offering ongoing professional development for everyone.

## PREPARING TO INITIATE PROJECT RAMP

IN 1998, as our district faced new mathematics program adoptions, we engaged some of our more reform-minded mathematics teachers in important conversations about the direction of the district's mathematics instruction for the coming five years and beyond. We were fortunate to have a cadre of teachers who had participated in a mathematics leadership initiative sponsored by the Education Development Center with NSF support. The Leadership in Urban Mathematics Reform (LUMR) Project teachers were willing to work with me to define a vision for long-term improvement in mathematics instruction. In addition to the LUMR teachers, several teacher leaders from our elementary schools were invited to join us for these planning sessions. In particular, as we examined the disparity in achievement in algebra 1 among various groups of students, we became determined to pave the way for more students to have access to challenging, meaningful mathematics. We decided that this goal could be met best through the use of integrated mathematics curricula. Therefore, we began promoting among our district's teachers and administrators the idea of adopting integrated mathematics materials. As a result, one year before specific program adoptions were to take place, the district decided to limit adoption choices to the thirteen integrated mathematics projects funded by the NSF.

We purchased sample sets of these projects, and during the adoption year, these materials along with reflection sheets traveled to each school.

This time period is now known in RAMP history as the time of the "traveling books." Some teachers decided to pilot-test some of the materials in their classrooms. Throughout the year, teachers and administrators spent a great deal of time evaluating the appropriateness of each of the various programs for use with our students. The adoption process was extensive—in retrospect, perhaps too extensive. As with any districtwide decision, not all stakeholders got the program they desired; however, the process was open, and the mission was clear. With excitement and trepidation, we began planning to implement Standards-based materials the next year.

## CHALLENGES AND OPPORTUNITIES

DURING THESE first years of Project RAMP, we encountered several challenges. We chose to view these challenges as opportunities to focus our work and refine our methods for improving achievement in students' mathematics performance. We faced the challenges of (*a*) the state's ABC accountability program, (*b*) furnishing professional development for 980 teachers, and (*c*) changing the mathematics curriculum.

## CHALLENGES OF THE ABC STATE ACCOUNTABILITY PROGRAM

OUR STATE'S accountability program is a high-stakes, reward-and-punishment system. Thus far, this program has presented four major challenges. First, because the EOG and EOC tests are based on multiple-choice items, the teachers worry about the discrepancy between teaching with open-ended investigations and testing with multiple-choice items. Second, when teachers and principals try to find a quick fix for lagging test scores, their efforts often result in a greater emphasis on getting good test results than on developing students' mathematical understanding. Third, the results from accountability testing can be deceiving because they do not take into consideration any weaknesses in schools' abilities to meet the educational needs of specific subgroups of students. Lastly, the high school EOC tests are based on a nonintegrated curriculum and, as such, do not align well with our integrated mathematics courses.

*Multiple-choice frenzy.* Although our state's curriculum and testing program are aligned, the state accountability system produces an environment that I call the "multiple-choice frenzy." The high-stakes, multiple-choice achievement tests are driving our work as they never have before. In response, the RAMP Project team produced support documents to show teachers that by teaching the integrated mathematics programs as designed, they are also teaching the state objectives and preparing students for the state's EOG and EOC tests. First, we correlated the state objectives with the Standards-based materials that teachers are using and created systemwide pacing charts for each mathematics course. Second, we created multiple-choice tests to be administered to students as benchmark tests every nine weeks. Each test is carefully aligned with the integrated mathematics program, state objectives, and pacing guides for the particular quarter. In these ways, we encouraged teachers to use the Standards-based materials even in the face of the multiple-choice frenzy.

*Desire for a quick fix.* Another challenge of the state accountability program is that it breeds the desire for a quick fix. The high-stakes nature of the testing system promotes an infatuation with workbooks and programs that offer a "teacher proof" solution to mathematics teaching. This infatuation is in direct opposition to the teaching of challenging, meaningful mathematics. The goal of Project RAMP is to give teachers long-term, sustained professional development opportunities, but this goal is undermined if principals convey the message that increasing test scores is all that matters. As an example of this concern, consider the following experience shared with us by one of our middle-grades teachers:

> After a recent workshop on the development of statistical teaching sequences, I presented my class with the battery data that we, as teachers, had explored during the workshop. The data was length of operation time before a battery went dead. Ten batteries each of two brands had been tested. Based on the spread of operation times for each brand, students were asked to determine which brand they would choose and to explain their choice. From the workshop, I understood the mathematical richness of the activity, and I was prepared to ask students probing questions to challenge their mathematical thinking. My principal called me outside during the lesson and challenged me with the question, "You need to be preparing your students for EOGs; why are you wasting valuable mathematics instruction time talking about batteries?"

Admittedly, principals carry a lion's share of the responsibility for ensuring students' progress in our state's high-stakes accountability program. Throughout our system, principals vary in their willingness to take risks, and not surprisingly, many are seduced by the idea of a quick fix. To counteract this tendency, we have encouraged principals to make their schools safe places for teachers to experiment with mathematics pedagogy as the teachers develop expertise in implementing Standards-based curricula.

*Exemplary growth for a few.* A third challenge that results from the high-stakes accountability system is the danger that a school can be designated an exemplary growth school on the basis of an increase in average performance even while specific groups of students do not display exemplary growth. This discrepancy particularly concerns me at the high school level, where the school average may be misleading because of the way that students enroll in courses. This situation can create a false sense of complacency. Although our district's testing office disaggregates achievement data by race, gender, and socioeconomic status, the state accountability system does not reward schools for the growth in achievement of separate groups of students. Likewise, the state accountability system does not specifically reward schools for increasing the number of students in higher-level mathematics courses at the high school level.

In contrast, the RAMP program closely monitors the patterns of students' enrollment in high school mathematics courses. Our minimum mathematics requirement for high school graduation is three units. In the past, less-advanced students could opt for courses in introductory mathematics, algebra 1A (the first half of an algebra 1 course), and algebra 1B (the second half of an algebra 1 course) to meet the graduation requirement. In essence, the combination of these three courses constituted no more than one year of high school level mathematics. With the integrated mathematics curriculum, the same students are expected to complete integrated mathematics courses 1, 2, and 3 as their three units of high school mathematics. Combined, these integrated courses are roughly equivalent to algebra 1, geometry, and algebra 2—three real units of high school mathematics. We clearly want to "raise the bar" of mathematics learning. However, by so doing, we are placing into our EOC testing pools students who, in the past, would never have taken these courses. At this point, we do not know what is going to happen. We face the possibility that even if students are learning much more in integrated courses than in the traditional ones, our overall EOC scores will drop simply because a greater number of "low-testing" students take the test. Should that happen, we may find ourselves in the position of having implemented a highly successful high school curriculum that gets condemned and tossed out because the test scores dropped. In Project RAMP, we are monitoring course-taking patterns, EOC scores, and dropout rates. Preliminary data strongly support continuation of integrated mathematics. We are also looking into ways of monitoring changes in students' attitudes toward mathematics. We believe that a combination of these outcomes offers the best measure of how successful the integrated mathematics program is for our students.

*Matching the high school curricula to the state tests.* A fourth challenge is that the timing and content of the state's high school tests do not match the sequencing of mathematics content in the integrated mathematics curriculum. For example, the state's testing program is based, in part, on EOC tests for algebra 1, geometry, and algebra 2. As the term EOC implies, these tests were designed to be administered at the end of an algebra 1 course, a geometry course, and an algebra 2 course, respectively. The state has not developed integrated mathematics 1, 2, and 3 EOC tests. Therefore, we have had to negotiate with the state to develop a testing schedule for students in integrated mathematics courses. We currently give the EOC test for algebra 1 at the end of integrated mathematics 2, the geometry EOC test at the midpoint of integrated mathematics 3, and the algebra 2 EOC at the end of integrated mathematics 3. This timing of the tests is needed because of the sequencing of the content in the integrated materials. Unfortunately, as a by-product of this testing schedule, we constantly have to battle the misconception that integrated mathematics 1 is a "slow" mathematics course because the students take the algebra 1 EOC test a year later than "normal."

## CHALLENGES OF PROVIDING PROFESSIONAL DEVELOPMENT

EACH OF the 980 mathematics teachers within our district are expected to participate in more than 100 hours of professional development in mathematics content and pedagogy over a four-year period. The sheer magnitude of Project RAMP's professional development component is daunting. As a district, we are facing many challenges in delivering the high-quality, long-term, sustained professional development needed to ensure that our vision of mathematics instruction is realized. These challenges include (*a*) competition for teachers' time, especially at the elementary school level, (*b*) lack of qualified elementary school mathematics professional development leaders, (*c*) teacher turnover, and (*d*) in some cases, resistance—particularly among high school teachers not teaching integrated mathematics—to making changes in their teaching practices.

*Competition for teachers' time.* We are all familiar with the myriad responsibilities that require teachers' time and attention. In particular, elementary school teachers face pressure to focus the instructional day around reading and writing. To overcome the difficulty of getting elementary school educators to focus on mathematics, the RAMP team has persistently trumpeted the cause of mathematics in administrative and school-based decision-making meetings. At any particular point in time, several professional development initiatives across different subject areas are typically competing for elementary school teachers' time. To ease the stress on teachers and to establish collaborative efforts with district administrative personnel responsible for ensuring high-quality teaching in other subject areas, the RAMP team constantly searches for ways to connect mathematics teaching with the teaching of other subject areas and makes a point of highlighting these connections.

With respect to professional development opportunities, we try to schedule as many of these sessions as possible during the professional development days set aside in our academic calendar. However, on those days when we are competing for time slots with all the other subject-area sessions within the district, this scheduling competition is especially problematic for elementary school and special education teachers. We have tried to be creative, flexible, and responsive to teachers' requests regarding the scheduling of additional professional development sessions. We have offered after-school and Saturday sessions and numerous summer sessions within the district. A few sessions have been offered during the school day, with the RAMP Project paying for substitute teachers. Especially at the elementary school level, we frequently offer a given professional development opportunity at a variety of times to accommodate teachers' schedules. We schedule all sessions well in advance and post the offerings on our district's professional development Web site. We encourage teachers to register for sessions online. Teachers receive a stipend for attending sessions outside their scheduled work hours and occasionally receive additional teaching materials (e.g., manipulatives and reference books) when they attend sessions. The RAMP Project staff documents the hours of professional development for each teacher and encourages teachers to complete the required number of hours.

School-based study groups, meeting once a month, have offered some opportunities for teachers to collaborate and prepare to implement new units within the integrated mathematics curricula. Unfortunately, these meetings often occur before or after the regular school day. Therefore, the study-group meetings fall short of addressing teachers' needs for collaborative planning time during the school day.

*Lack of qualified elementary school leaders.* A second challenge that the RAMP project faced as we implemented staff development was the lack of "homegrown" mathematics leaders at the elementary school level. We decided early on that it was best that all teachers begin professional development on the integrated mathematics programs at the same time. We also believed that it was important to have RAMP teacher leaders available at every school. These two conditions required that we prepare approximately sixty elementary school teacher leaders. Although some of the elementary school teachers that we selected were ready for this role, many of them were not. To encourage and assist those in the latter group, we paired experienced workshop leaders with leaders needing assistance and developed extensive workshop leader notes. As we entered the second year, we varied our schedule to reduce our need for districtwide leaders. This modification has improved the quality of the workshops but has created new problems with regard to tracking workshop scheduling and attendance.

*Teacher turnover.* A third challenge—teacher turnover—affects the RAMP project at all grade levels. We have invested many workshop and coaching hours with a teacher only to find out that the teacher was leaving the district. To address continuity problems when teachers leave, we are attempting to offer workshops on a continuous cycle. This schedule is difficult when most of the workshops that we offer must be designed "from scratch." We also address turnover issues by working with the district's human resources department to stay apprised of staff changes. This knowledge allows us to at least visit new teachers and help orient them to the materials.

*High school teachers who do not teach integrated mathematics.* At the elementary and middle school levels, all mathematics teachers use *Investigations* or *Mathscape: Seeing and Thinking Mathematically* to teach integrated mathematics. At the high school level, our integrated mathematics program is offered alongside the traditional sequence of algebra 1, geometry, and algebra 2. Although our elementary and middle school teachers are at various stages in implementing the curriculum, the professional development program for them can easily revolve around the common curriculum and mathematics content. At the high school level, the teachers who do not teach integrated mathematics are still involved in professional development along with the teachers who do. At first we struggled with how to design appropriate offerings for these teachers. During the first year, we pieced together a menu of professional development offerings that we believed might help teachers of the more traditional course sequence become more reform-minded. During the second year, we chose the *Data Driven Mathematics* series (Burrill et al. 1999) to guide the professional development sessions for teachers who were not teaching integrated mathematics. During these sessions, the teachers experience the highly integrated units *as learners* and *as teachers* of mathematics. The sessions give us a chance to show teachers the potential of investigative, integrated instructional units to help students develop deeper understandings of mathematics content. We have also used various resources from our state mathematics office. With these approaches, we can offer offering meaningful professional development to all the mathematics teachers in our district.

## CHALLENGES OF CHANGING THE CURRICULUM

MOVING TO an integrated mathematics curriculum for kindergarten through eighth grade and a choice of an integrated mathematics curriculum for ninth through twelfth grades presents several challenges to our district. These challenges include (*a*) phasing in the curriculum, (*b*) competing for instructional time in kindergarten through fifth grade, (*c*) overcoming the hesitancy of high school teachers to accept and implement the integrated mathematics sequence of courses, and (*d*) smoothing students' transition between middle and high school mathematics.

*Phasing in the curriculum.* Our district chose to phase in the elementary and middle school curricula by units rather than teach all the units in the first year. Although we have addressed all the major implementation obstacles (e.g., obtaining materials, conducting professional development sessions, and designing pacing charts and assessments) and are ready for full implementation, some of our teachers are simply not implementing the curriculum as scheduled. Rather than teach the full units, some teachers pick and choose activities from the integrated mathematics materials; others pull activities from a variety of other instructional resources. While honoring their needs and capacity for change, we are attempting to nudge all teachers toward fully implementing the integrated mathematics materials. We are working with school-based teams to write the integrated mathematics materials into their school improvement plans. We are also offering such incentives as extra student materials and more coaching hours to schools that are ready to fully implement the program. We believe that the power of the curricula will never be realized unless the teachers use the entire curricula as intended.

Phasing in integrated mathematics at the high school level presents its own challenge. The *Contemporary Mathematics in Context* integrated curriculum is being phased in by course. We planned for integrated mathematics 1 to begin in the first year and integrated mathematics 2 to be introduced in the second year. Integrated mathematics 3 and integrated mathematics 4 were planned to be introduced in the third and fourth years, respectively. However, in reality, only four of the six high schools began offering integrated mathematics 1 in the first year; one school began in the second year, and the other school is planning its first integrated mathematics offering in the third year. This course staggering makes quite challenging the task of placing students transferring from one school to another within the district. In such situations, we work out individual plans for students as needed.

*Competing for instructional time.* Teaching the integrated mathematics curricula requires more time for teachers and for students; this challenge affects mainly the elementary school level. Having embraced a reading-and-writing model that takes all the morning teaching time at most elementary schools, our district has several schools that struggle to offer a full hour of mathematics each day—even though that is our district standard. We address this challenge by furnishing data and reminders to principals regarding the district standard. We also work with our colleagues in reading, writing, science, and social studies to find links enabling all subject areas to be taught together in a fully integrated unit. Clearly, our "secret weapon" for increasing mathematics achievement is to teach mathematics, and teaching the mathematics curriculum requires time.

*Hesitancy of high school teachers.* A third challenge is high school teachers' caution toward, and sometimes reluctance to accept, our integrated mathematics program as a legitimate sequence of courses. Integrated mathematics is perhaps most daunting at the high school level. The appearance and organization of the curriculum materials invite a comparison with the familiar sequence of algebra 1, geometry, and algebra 2 that the high school mathematics teachers have known for a long time. One teacher said, "I will quit before I will use materials like this." I have had others admit that they are skeptical, but they sincerely and fairly delve into learning the content and the pedagogy that is inherent in the integrated mathematics program. Although teachers exhibit various levels of acceptance, most of those who are teaching integrated mathematics have quickly committed to the materials and the approach. These high school teacher leaders provide an excellent service to the district by promoting the program.

In general, the RAMP Project has tried to invite rather than force teachers to use integrated mathematics materials. Principals have been influential and successful in recruiting the next group of integrated mathematics teachers as we have needed them. We spend a great deal of time explaining what integrated mathematics is and what it is not. We have written a brochure about integrated mathematics. We speak about integrated mathematics to eighth-grade teachers, eighth-grade counselors, high school teachers, high school counselors, parents, principals, and whoever else will listen. Counselors are particularly influential in helping students make decisions about mathematics courses. We are ever mindful that all our work could be lost if we lose the trust of the community. We began this sequence of courses to offer a choice for students and to reach our goal of meaningful mathematics for all students.

*Smoothing the transition.* A fourth challenge is students' transition from eighth to ninth grade. Although a few of our middle schools offer one or two years of integrated mathematics, most of our middle schools offer first-year algebra to eighth-grade students who are one year ahead of their peers in mathematics. We want these students to have the choice of integrated mathematics as they enter high school. To meet this challenge, we allow rising ninth-grade students who have already completed algebra 1 to go directly into integrated mathematics 2. In this way, we make sure that integrated mathematics is seen as an option for all students as they enter high school. A student who completes integrated mathematics 2 in ninth grade would be expected to complete three more units of mathematics (presumably integrated mathematics 3, integrated mathematics 4, and AP Calculus) to the meet the graduation requirement of four mathematics units.

## ADVOCACY WITH STAKEHOLDER GROUPS

A VITALLY important facet of my work is advocacy with various stakeholder groups. I am amazed at the amount of time and energy that goes into building and sustaining change. My mindset as the mathematics coordinator for this district is that all of us—teachers, administrators, students, parents, and community members—can learn, and all of us will be needed to reach all the students. We must keep searching for challenges and ways to overcome them. One of my most difficult tasks is relating to those who criticize the direction in which our mathematics instruction is headed. I find that my greatest tool is the students' achievement data. On a regular basis and with different stakeholder groups, I *gently* encourage all to carefully consider who is and who is not achieving in mathematics.

## ADVOCACY WITH TEACHERS

I MUST honor school culture and teacher readiness for change while presenting a vision of what improvements are possible. My approach has been to develop teacher leaders and give them multiple chances for professional development in mathematics and leadership. Teacher leaders are essential to the approach that we are using to implement Standards-based materials. Those leaders are the ones who are involved in the day-to-day decisions and the school-level work involving curriculum, assessment, budgets, and parental involvement.

## ADVOCACY WITH PRINCIPALS, ADMINISTRATORS, AND THE SCHOOL BOARD

IN ADDITION to teachers, I must reach out to principals and administrators. I have identified such existing communications avenues as monthly principals' meetings and the principals' Web site as ways to keep administrators informed and solicit their support. The RAMP team prepared a notebook on the integrated mathematics programs and encouraged teacher leaders to work with their principals to craft a vision for mathematics at their individual schools. Principals have also asked for such public relations items as brochures, newsletters, and videos—all of which assist them in communicating with the community. The RAMP team continues to develop these items in collaboration with the schools.

In addition to teachers and principals, the RAMP team and I must advocate for the Standards-based programs with our executive-level administrators and our school board. It is important that the district leaders acknowledge the implementation of the Standards-based materials as part of the district's overall vision for improving instruction. Our district's overall vision for quality instruction is heavily influenced by Phil Schlechty's Center for Leadership Reform and its efforts in our district. The work of RAMP fits easily into the framework for reform articulated by Phil Schlechty (Schlechty 2001). To facilitate a connection between RAMP and the district's vision for quality instruction, the RAMP team connects the Standards-based materials with the articulated vision of our district whenever the opportunity arises. RAMP leaders are team players who also trumpet the cause of improving mathematics education. We also keep executive-level administrators and the school board informed about national and international trends and data regarding mathematics achievement, and we furnish periodic updates on the progress toward our vision.

Another group that has been fundamental in maintaining our momentum toward better mathematics teaching has been my colleagues in the central administration. As the mathematics coordinator, I search for connections among ongoing district initiatives. I not only work hard to keep connected with initiatives in other subject areas but also look for connections with those who lead our district in guidance, dropout prevention, community outreach, testing and accountability, public affairs, and technology. For example, because we have worked together with eighth-grade and ninth-grade guidance counselors, we are finding that many more students are choosing high school integrated mathematics courses. In addition, the support of the guidance counselors has been indispensable in the effort to introduce integrated mathematics to parents and the community.

## Advocacy with Parents and the Community

BECAUSE PEOPLE in our community do care deeply about who is and who is not succeeding in mathematics, the RAMP team is open and inviting to parents and the wider community regarding our mission. In general, humility is very important in the approach to the community. Our approach involves giving community members access to the curriculum and offering workshops in which we model effective ways to help and encourage children in mathematics. We are striving to give people in the community a chance to learn the mathematics and see themselves as teachers of their children. When disagreements or concerns exist, we seek to listen carefully and respond respectfully. We find such places of agreement as the need for computational fluency while inviting the community members to expand their notion of basic skills to include reasoning and problem solving.

## Recommendations and Concluding Remarks

AS A district, we began this journey by drafting a vision of mathematics instruction that we believed would serve all our students. This vision was defined by a select group of teachers and teacher leaders within our district. This group then promoted the defined vision throughout the district. At times, we have lamented that stakeholders who do not share our vision have made districtwide implementation of integrated mathematics much more difficult. When I find myself thinking this way, however, I remember the notion of "taken as shared" put forth by Paul Cobb (Cobb 1991, 2000). Taken-as-shared describes the understanding that develops when a class engages in investigative problem solving and discourse about the mathematical ideas of individual students. Taken as shared also describes the collaborative interpretations of a classroom teacher and an educational researcher regarding events and processes that occur in the classroom.

At the district level, I think something analogous to taken-as-shared understanding must develop for all stakeholders to buy into and promote a common vision for mathematics instruction. We really do not want the move to a common vision to happen too easily. We want teachers to have opportunities to engage in important conversations with their colleagues about how students' mathematics achievement varies with changes in instructional mode. We want teachers to grow in their abilities to attend to students' explanations and assess students' mathematics understanding. And we want other stakeholders to reflect on what they understand about fostering mathematical knowledge and how our students might be best helped to learn. Creating such a taken-as-shared vision, if it happens, would be the ultimate success story of the RAMP Project.

Districts that choose to implement Standards-based materials would be wise to plan for long-term, sustained professional development and to maintain their focus on the success of all students. We recommend a data-driven, thoughtful process for laying the groundwork for change. We also recommend facing challenges as opportunities and doing the time-intensive work of advocacy with the various stakeholder groups. We have only begun our journey, and we are hopeful and determined that Standards-based materials will assist us in achieving our vision of challenging, meaningful mathematics for all. So the bottom line in my advice to other districts that are embarking on a similar journey is to never lose sight of the vision, never miss an opportunity to promote that vision, and be gentle with the stakeholders as they struggle to reconcile their experiences with that vision.

## REFERENCES

Burrill, Gail F., Jack Burrill, Miriam Clifford, and Emily Errthum. *Data-Driven Mathematics.* Palo Alto, Calif.: Dale Seymour Publications, 1999.

Cobb, Paul. "Reconstructing Elementary School Mathematics." *Focus on Learning Problems in Mathematics* 13(2) (1991): 3–22.

———. "Conducting Teaching Experiments in Collaboration with Teachers." In *Handbook of Research Design in Mathematics and Science Education,* edited by Anthony E. Kelly and Richard A. Lesh, pp. 307–34. Mahwah, N.J.: Lawrence Erlbaum Associates, 2000.

Core-Plus Mathematics Project. *Contemporary Mathematics in Context: A Unified Approach.* Columbus, Ohio: Glencoe/McGraw-Hill, 1998.

Education Development Center. *Mathscape: Seeing and Thinking Mathematically.* Columbus, Ohio: Glencoe/McGraw-Hill, 1998.

National Council of Teachers of Mathematics (NCTM). *Curriculum and Evaluation Standards for School Mathematics.* Reston, Va.: NCTM, 1989.

———. *Principles and Standards for School Mathematics.* Reston, Va.: NCTM, 2000.

North Carolina Department of Public Instruction, Division of Accountability Services. 2000–2001 ABCs Results. [cited 9 January 2003].Available at http://www.ncpublicschools.org/abc_results/results_01/leas/320.html; World Wide Web.

Schlechty, Phillip C. *Shaking Up the Schoolhouse : How to Support and Sustain Educational Innovation.* 1st ed. San Francisco: Jossey-Bass, 2001.

TERC. *Investigations in Number, Data, and Space, Grades K–5.* Glenview, Ill.: Scott Foresman Educational Publishers, 1997.

20

# Implementing Integrated Mathematics Programs: The Vision

Sharon Stenglein

THE VISION for school mathematics education described in *Principles and Standards for School Mathematics* and excerpted below is highly ambitious (NCTM 2000, p. 3):

> Imagine a classroom, a school, or a school district where all students have access to high-quality engaging mathematics instruction.... The curriculum is mathematically rich, offering students opportunities to learn important mathematical concepts and procedures with understanding. Technology is an essential component of the environment. Students confidently engage in complex mathematical tasks chosen carefully by teachers. They draw on knowledge from a wide variety of mathematical topics, sometimes approaching the same problem from different mathematical perspectives or representing the mathematics in different ways until they find methods that enable them to make progress.... Students are flexible and resourceful problem solvers. Alone or in groups and with access to technology, they work productively and reflectively, with the skilled guidance of their teachers.

This new vision is one shared by many of those involved in mathematics education. And implied in this vision, although not explicitly stated, is the importance of integrating mathematical ideas.

## THE MEANING OF INTEGRATED MATHEMATICS

AN OVERALL issue related to integrated mathematics involves defining the term *integrated.* An integrated program might feature weaving together various strands of mathematics content within a course, connecting mathematics with other subjects, or using mathematics in real contexts. Occasionally the term integrated is even used as a moniker for any mathematics program that differs from what most adults remember from their own schooling. As it is used in this chapter, however, integrated mathematics means a mathematics program that—

- is aligned with the vision and Standards for school mathematics set forth by the National Council of Teachers of Mathematics;
- is mathematically rich and challenging and contains engaging tasks drawn from multiple strands of mathematics;
- is accessible to all students;
- is coherent, with a nonrepetitive scope and sequence across several years or courses;
- includes research-based, effective instructional practices that help students make sense of mathematics and develop mathematical reasoning skills; and
- encourages mathematical interactions within a learning community of students and teachers.

Thus integrated mathematics is not (*a*) simply teaching from a textbook that includes chapters from various mathematics content strands, (*b*) using a thematic approach, (*c*) coupling the teaching of mathematics with another subject, or (*d*) teaching a single unit or course that blends parts of mathematics or mathematics and other subjects. Unless all the components in the bulleted list above are present, the study of mathematics and mathematics programs will not be considered integrated for the purposes of this chapter.

## RATIONALE FOR INTEGRATED MATHEMATICS

A GROWING body of evidence suggests that the compartmentalized approach of the past has not been successful in supporting students' learning of powerful mathematics. For example, more than half the enrollment in college mathematics is in classes ordinarily taught at the high school level (Battista 1999). Furthermore, research shows that students in a traditional, textbook-based program (in which teachers present a series of procedures arranged by separate mathematics topics) are outperformed in nearly all measures by students engaged in solving problems that are not segmented by mathematical strand or topic (Boaler 1997).

Because many programs that fit the definition of integrated mathematics outlined above are quite new, longitudinal data on their effectiveness does not yet exist. Nevertheless, preliminary data from many of these new integrated programs, results from cognitive science about what drives effective learning of mathematics, and data about the segmented programs that we have been using suggest great potential for increased student achievement in integrated mathematics programs (Battista 1999).

## MEETING THE CHALLENGES OF INTEGRATED MATHEMATICS: ROLES AND RESPONSIBILITIES

SERIOUS CHALLENGES face those who are trying to implement integrated mathematics programs. The remainder of this chapter examines these challenges in relation to the roles and responsibilities of the various stakeholders, starting with those at the center of the classroom and moving outward.

### Students of Mathematics

Students often will work hard if they are both challenged and supported to achieve success. They take their cues from one another and from adults. If students do not hear their peers, teachers, or parents or other caregivers criticize integrated programs, they are more likely to participate in such programs enthusiastically. In addition, students often find the study of integrated mathematics to be more engaging and meaningful than their study of more traditional mathematics was. Consequently, they are often motivated to extend their study of mathematics, even when doing so is elective. If high school students can choose between an integrated and a sequential program, their discontent is greatly reduced.

A particular challenge comes when students at all grade levels in a school begin a new integrated program at once. Students at upper grade levels face at least two disadvantages: (1) they are not beneficiaries of the scope and sequence of the integrated program, and (2) they are subject to classroom experiences different from those they have had previously. These students are asked, for example, to approach problems in multiple ways, to explain their thinking orally and in writing, and to work with other students. Students who have previously been very successful with minimal effort or who have been rewarded for speed and accuracy in computation may resist these changes. In a classroom situation in which students regularly share their thinking with one another, some students may become impatient with peers who think differently or more slowly. Only with time and new experiences do students realize the benefits that come from hearing multiple approaches to solving a problem.

A different issue arises if an integrated program is implemented one year or one course at a time. Each year, the same group of students becomes the pilot group for a new teacher using the program for the first time. One solution involves implementing one grade or course at a time and having the teachers move up with the students for two or three years. This progression is particularly recommended for the high school level, where it is imperative that an integrated program be implemented one year at a time. A systemwide integration program could be effectively launched in three to four phases. In the first year, integrated programs would be implemented at the levels of kindergarten, grade 3, grade 6, and grade 9. The following year, the program could add the subsequent four grades: grades 1, 4, 7, and 10; and the third and fourth years, the remaining grades would be added to complete the implementation. Another option at the elementary school level is to designate at each grade level teachers who will teach the mathematics in several different classrooms to give more students the benefit of their expertise.

## Teachers of Mathematics

Teachers are being asked to do nearly impossible tasks as they work to implement integrated mathematics programs: (*a*) teach content that is often new and unfamiliar to them, (*b*) make mathematical connections that are also new, and (*c*) use teaching strategies that they have never experienced as learners. Therefore, the training of teachers must go beyond an overview of the program, which was often sufficient in the past, and must include lessons modeled by experts and opportunities for teaching sample lessons and receiving feedback. Teachers implementing integrated mathematics programs also need ongoing mentoring and coaching from peers and leaders.

Instructional planning and performance assessment in an integrated program place large demands on teachers' time. Yet most teachers continue to work in isolation, with little of the needed support for innovation and few of the necessary incentives to improve their practice. Compounding the problem are the facts that a significant number of teachers of mathematics do not hold major or minor degrees in mathematics and that, in the past, some elementary school teachers have demonstrated limited in-depth understanding of mathematics (Ma 1999). Furthermore, in a recent survey of U.S. teachers, the data illustrate the minimal mathematics content preparation of teachers in kindergarten through grade 6. Although nearly all elementary-level teachers have taken both a mathematics course for elementary-level teachers and a mathematics education (methods) course, only 38 percent have taken a college algebra course and only 31 percent have had a course in probability and statistics. Only one elementary school teacher in five has taken a geometry course and only one in ten has studied any calculus. (Weiss et al. 2001) Even where qualified teachers are available, serious questions surround the equitable distribution of those teachers among schools (Kilpatrick and Silver 2000). In integrated mathematics programs, teachers find a compelling need—more so than they had found previously in traditional programs—to collaborate with colleagues to share ideas and plan, review, and revise lessons. This collaboration is both a support and a more efficient way to work.

Implementing integrated mathematics programs poses multiple challenges for teachers, and meeting those challenges requires support from many sources. Teachers need high-quality staff development opportunities, ongoing and sustained support and leadership, resources to support instruction, and creative scheduling to allow time for planning and collaboration. Teachers' own responsibilities in meeting these challenges include becoming proactive in making their professional needs known and taking advantage of available opportunities.

Additional implementation issues for teachers pertain to their beliefs and the quality of instruction. Teachers have strong beliefs about what mathematics is, who can learn it, and how it is best learned. New programs often challenge some of these beliefs and require all educators to expand their ideas about the teaching and learning of mathematics. Issues of instructional quality arise when teachers who have been successful in more-structured teaching situations are required to help students learn mathematics in new ways. Whether because of lack of confidence, experience, or staff development and support or because of other personal limitations, some teachers find this transition to be overwhelming. Implementing an integrated mathematics program creates a greater range of quality in mathematics instruction. The challenge of maintaining the quality of instruction also affects the roles and responsibilities of administrators, supervisors, staff developers, and teacher educators and must be faced head on.

Other responsibilities for teachers include helping students, especially high school students, be aware of how to describe in traditional terms the mathematics they know. When students complete questionnaires of prior coursework, they will need help in identifying their knowledge of algebra, geometry, statistics, and so forth. Secondary school mathematics specialists will need to pursue professional development in strategies for teaching reading because contextualized problems often require extensive reading by students. A bonus derived from teaching integrated mathematics is the job-embedded growth that comes from learning mathematics with one's students. Teachers can also engage in formal or informal classroom research as they monitor the effect of the program on students and listen to students describe their thinking to evaluate their learning. Teachers must plan carefully for instruction, preparing questions that will elicit mathematical thinking and ensuring that the program offers entry for all students and also "residue" for all (Hiebert et al. 1997). Teachers should learn to trust programs that have been developed by experts in mathematics and learning and refrain from reverting to more familiar activities and exercises. Finally, but foremost, teachers of integrated mathematics must always serve as the keepers of the mathematics for their students by bringing closure to lessons and not getting lost in the application or context at the expense of the mathematics.

## School Administrators and Staff

School administrators need to be knowledgeable about the integrated mathematics programs used in their schools and the new roles that the programs require of the teachers. Administrators must be involved throughout the program-selection process (where that is a local decision) and have experienced sample lessons to help them understand the intent and structure of the program. In addition, administrators themselves have multiple ongoing roles and responsibilities. Their willingness to hear and answer questions from community members regarding integrated mathematics programs, even if those questions are skeptical of the program, frees teachers to focus on planning and collaboration. Administrators' support of the program with time and resources is fundamental to its success. For example, principals' willingness to adjust the schedule can give teachers essential time for collaboration, and principals' ability to set priorities and allocate resources can support essential professional development. In their role as evaluators of teachers' performance, administrators must ensure that their evaluations reflect both the vision of mathematics teaching and learning in *Principles and Standards for School Mathematics* (NCTM 2000) and each teacher's ability to skillfully create a mathematical learning community, even if that community does not look like a model classroom of the past. This evaluation also needs to include coaching for improving professional practice along with whatever resources are needed for that improvement.

Counselors, especially at the high school level, play an important role. In conjunction with administrators, counselors must allow each student to participate in courses that will best suit the students' learning goals and needs. This placement of students requires that counselors be well informed about, and supportive of, integrated mathematics programs. High school counselors have the primary responsibility in clearly articulating on students' transcripts the mathematics content and learning so that colleges and employers can recognize it. They also are responsible for evaluating and placing students transferring into a school that uses an integrated program. Counselors need to dispel the notions that integrated programs are not suitable for college-bound students and that the mathematics in these programs is not as rigorous as the mathematics in a traditional program. Counteracting these myths is especially crucial where both integrated and traditional programs are taught in a high school. Counselors should also help collect data regarding the performance of their integrated mathematics students on assessments, including college-entrance tests and state and local assessments.

## Teacher Leaders and District Mathematics Supervisors

Mathematics teacher leaders in a school and mathematics supervisors in a school or district have ongoing roles and responsibilities in structuring integrated mathematics programs and supporting teachers. These leaders must ensure high-quality staff development for all by working with those who develop and present professional development programs. High-quality professional development that occurs consistently and over time will help entice reluctant teachers to participate. Furthermore, teacher leaders and supervisors can plan and lead study groups for teachers and give teachers opportunities to plan together. Teacher leaders and supervisors can support time for teachers to (*a*) collaboratively refine lessons, (*b*) visit one another's classrooms, (*c*) collectively examine students' work and suggest further instruction, and (*d*) make sound decisions regarding content that reflects local or state priorities. Teacher leaders and supervisors should mentor teachers, both new and experienced, and encourage reflective practice among teachers of mathematics. Although these professional support activities are always desirable, they are imperative for the success of integrated programs.

Supervisors can also support teachers by planning opportunities for parents and community members to learn about integrated mathematics programs. Such opportunities include parent meetings, open houses, newsletters, and presentations at community events. These meetings present opportunities to show parents the differences between programs. For example, one approach involves asking adults to remember their own mathematics experiences. Next a teacher can share from the integrated mathematics program a lesson that participants can contrast with their own prior experience. Another related duty of a supervisor is to meet with students and parents prior to transition points in the system. They should discuss the mathematics program at the next level and any choices that the parents and students may need to make. For example, if a high school offers both an integrated and a traditional program, middle school students and parents should understand the strengths of the integrated program and have an opportunity to experience one or more integrated mathematics lessons. Supervisors can also collaborate with district assessment personnel in collecting local data on students' achievement. The support and leadership of teacher leaders and supervisors is essential for effectively implementing an integrated mathematics program.

## Curriculum Committees and Curriculum Coordinators

The role of a curriculum committee is extremely important for implementing a high-quality mathematics program. The committee's role, however, varies widely depending on whether programs are selected at the state, district, or site level. Regardless of the locus of control, the responsibility of the committee or group is serious. The committee must represent the larger group of stakeholders and must include parents who have credibility in the community. The committee members must be enlightened and aware of the steps in the process of change. They must know the requirements of the state, district, or school they represent and the resources available and necessary for implementation—particularly professional development resources. Committee members must listen to, but not be overly influenced by, special interest groups. They must be willing, especially if they are working at the local level, to slow down or extend the transition process to build support from a critical mass of educators and the public. Special issues that these curriculum decision makers face include determining the sequence of implementation and, at the secondary school level, whether to offer integrated and sequential mathematics programs side-by-side to give students and parents a choice. Furthermore, the curriculum committee must inform the public about the criteria and processes used in decision making and build support for change.

The public should understand that because a conceptually based integrated program helps students permanently grasp content, with such a program it becomes unnecessary to repeat topics each year. Articulation and planning are particularly necessary at the grades in which students make a transition from an elementary school program to a middle school program and from a middle school program to a high school program. To ensure that no gaps or unnecessary overlaps exist between levels, teachers experienced in each program must meet to negotiate who will teach what units so that they offer a seamless program for students. The curriculum coordinator—either a mathematics specialist or a generalist—has particular responsibility for leading and structuring the committee to ensure that their work will be both productive and representative.

## District Administrators and School Boards

Because district administrators and school board members are responsible for all students in all learning areas, their concerns extend beyond mathematics programs. However, their roles related to mathematics programs are complex and important. Of particular concern to district administrators and school boards are the structure of the district's schools, the allocation of resources for materials and professional development, articulation among grade levels, and policies related to curriculum and staff development. They must devise structures and policies that encourage and reward teachers for participating in professional development opportunities that improve teaching and learning.

The grade-level configuration of school buildings poses particular challenges for implementing integrated mathematics programs. For example, many integrated mathematics programs at the secondary school level begin at grade 9. If a district has a junior high school for grades 7–9 and a senior high school for grades 10–12, collaboration among teachers is more difficult. Often those who are teaching the courses for tenth-grade students are not familiar with the content experienced by their students the previous year. This dichotomy severely compromises the scope and sequence of an integrated program. Multiage classrooms can pose another challenge. In situations where classes include students grouped by various ages rather than by learning needs, implementing a program with a scope and sequence is extremely challenging. Although we would be unrealistic to expect district administrators and school boards to make decisions based solely on the needs of a mathematics program, those needs must be a component of their decisions if students' learning needs are to be served.

For their decisions and responsibilities to reflect actual practice rather than hearsay, district administrators and school board members must also be well informed about, and have firsthand experience in, integrated mathematics programs. These leaders allocate resources and set policies related to articulation among various levels. In doing so, they must support program planning from kindergarten through grade 12 and sustained, high-quality staff development for all staff members, including such support staff as those working with special education students and English-language learners. District policies and practices must encourage teachers' professionalism and reflective practice. When district administrators and school board members communicate with the public, they must explain rationales for new programs and plans for monitoring students' progress in integrated mathematics programs. Even districts with strong current achievement need to continue to improve in our rapidly changing society.

## Teacher Educators and Higher Education Faculty

Multiple responsibilities fall to the teacher education and mathematics faculty members in institutions of higher education. These educators prepare teachers for initial licensure, administer continuing professional development for teachers of mathematics in degree and nondegree programs, and have a vested interest in how well the system for kindergarten through grade 12 prepares students for higher education. All higher education faculty members must take seriously their responsibility to be well informed about integrated mathematics programs and about the implications of educational and cognitive research for teaching and learning mathematics. Putting this research into practice will give the teachers and the students they work with multiple opportunities to experience and practice teaching and learning in environments consistent with integrated mathematics programs.

To make the transition to higher education for today's students as smooth as possible, higher education faculty need to be knowledgeable about the content and methods used in integrated programs. They also must be willing to furnish valuable feedback to educators in kindergarten through grade 12 about the knowledge and skills of students who have graduated from integrated programs. College-level faculty members are in a unique position to judge whether more students are participating in advanced mathematics classes and whether those students are truly learning the content of those courses. These mathematics educators can offer ongoing data on the successes and weaknesses of integrated programs.

All higher education faculty members should continually evaluate such predictors of college success as entrance examinations and placement tests. When necessary, each measure must be revised—both as college courses change and as the skills and understandings of incoming students evolve. The attitude of college and university faculty toward integrated programs should reflect their openness toward continual improvement of mathematics programs. Their ability to reserve judgment of integrated programs while they await more longitudinal data will demonstrate their confidence in the structures already in place, for example, college entrance scores and department placement examination results, which currently predict which students are prepared for college coursework.

## State Mathematics Curriculum and Assessment Supervisors

The duties of state mathematics and curriculum specialists vary widely across states and are often performed by multiple individuals in a given state. In general, state mathematics supervisors must support educators who are implementing integrated mathematics programs and must work to establish state policies that align all parts of the system, especially assessment and frameworks, with such programs. For example, the state assessments should be consistent and aligned with the outcomes of an integrated mathematics program—a special concern in states with assessments in each grade or end-of-year tests in algebra or geometry. The end-of-year subject tests pose particular challenges because students in integrated mathematics programs study these strands over several years and may not be fully prepared at the time a traditional end-of-course examination is given.

Various issues surround curriculum selection in states with statewide adoption of instructional materials as well as those with more local control of curriculum. In the former, the state supervisor may be able to advocate including integrated mathematics programs among the choices. In states allowing more local control, the decisions on instructional materials are often made school by school, and supervisors must convey information on integrated mathematics programs to district and school decision makers. Supervisors often can ensure that professional development for those implementing integrated mathematics programs is of high quality and is sustained and that states allocate sufficient resources for it. Finally, state supervisors should collect and disseminate research on integrated mathematics programs gathered at the state, national, and international levels. Such research includes achievement data as well as information on the components necessary for successful implementation, the criteria for program selection, any in-depth evaluations done by national organizations, and the teaching practices that maximize students' learning. State supervisors can support educators implementing integrated mathematics programs by serving as clearinghouses and resources for districts, schools, and individuals.

## CONCLUSION

WITH AN integrated mathematics program in place, educators can arrive at the following vision for school mathematics, first posed by E. H. Moore in 1902, that has been evolving over the years:

> Thus the proposed four years' laboratory course in mathematics and physics will come into existence by way of evolution. In a large secondary school, the strongest teachers, finding the project desirable and feasible, will establish such a course alongside the present series of disconnected courses—and as time goes on their success will in the first place stimulate their colleagues to radical improvements of method under the present organization and finally to a complete reorganization of the courses in mathematics and physics. (Moore 1926, p. 54)

Integrated mathematics programs offer great opportunities for students to develop strong mathematics skills. Yet implementing such programs poses a formidable challenge to the educational community. The mathematics in these programs is not always obvious, and one needs to become a student again and participate in learning activities to fully grasp the often subtle but deep mathematics in a lesson. Only if all stakeholders accept the responsibility of their positions can the programs reach the highest levels of success. This success will happen gradually, over time, with both forward and backward steps. As students experience the rewards of learning mathematics that makes sense and realize the satisfaction and sense of empowerment associated with such learning, they themselves may become the best promoters of integrated mathematics programs. Increased achievement for students in schools that carefully implement integrated mathematics programs should win over parents, employers, higher education faculty, and community members. No one wants to argue with strong and increased student learning.

## BIBLIOGRAPHY

Battista, Michael T. "The Mathematical Miseducation of America's Youth: Ignoring Research and Scientific Study in Education." *Phi Delta Kappan* 80 (February 1999): 424–33.

Boaler, Jo. *Experiencing School Mathematics: Teaching Styles, Sex, and Setting.* Buckingham: Open University Press, 1997.

Hiebert, James, Thomas P. Carpenter, Elizabeth Fennema, Karen C. Fuson, Diana Wearne, Hanlie Murray, Alwyn Olivier, and Piet Human. *Making Sense: Teaching and Learning Mathematics with Understanding.* Portsmouth, N.H.: Heinemann, 1997.

Kilpatrick, Jeremy, and Edward A. Silver. "Unfinished Business: Challenges for Mathematics Educators in the Next Decades." In *Learning Mathematics for a New Century,* 2000 Yearbook of the National Council of Teachers of Mathematics (NCTM), edited by Maurice J. Burke, pp. 223–35. Reston, Va.: NCTM, 2000.

Ma, Liping. *Knowing and Teaching Elementary Mathematics: Teachers' Understanding of Fundamental Mathematics in China and the United States.* Mahwah, N.J.: Lawrence Erlbaum Associates, 1999.

Moore, Eliakim Hastings. "On the Foundations of Mathematics." In *Readings in the History of Mathematics Education,* edited by James K. Bidwell and Robert G. Clason, pp. 246–255. Washington, D.C.: National Council of Teachers of Mathematics, 1970. Originally given as a presidential address before the American Mathematical Society (29 December 1902).

National Council of Teachers of Mathematics (NCTM). *Curriculum and Evaluation Standards for School* Mathematics. Reston, Va.: NCTM, 1989.

———. *Professional Standards for Teaching Mathematics.* Reston, Va.: NCTM, 1991.

———. *Assessment Standards for School Mathematics.* Reston, Va.: NCTM, 1995.

———. *Principles and Standards for School Mathematics.* Reston, Va.: NCTM, 2000.

SciMathMN. *Minnesota K–12 Mathematics Framework.* Saint Paul, Minn.: SciMathMN, 1998.

Weiss, Iris R., Banilower, Eric R., McMahon, Kelly C., and Smith, P. Sean. *Report of the 2000 National Survey of Science and Mathematics Education.* Chapel Hill, N.C.: Horizon Research, 2001.